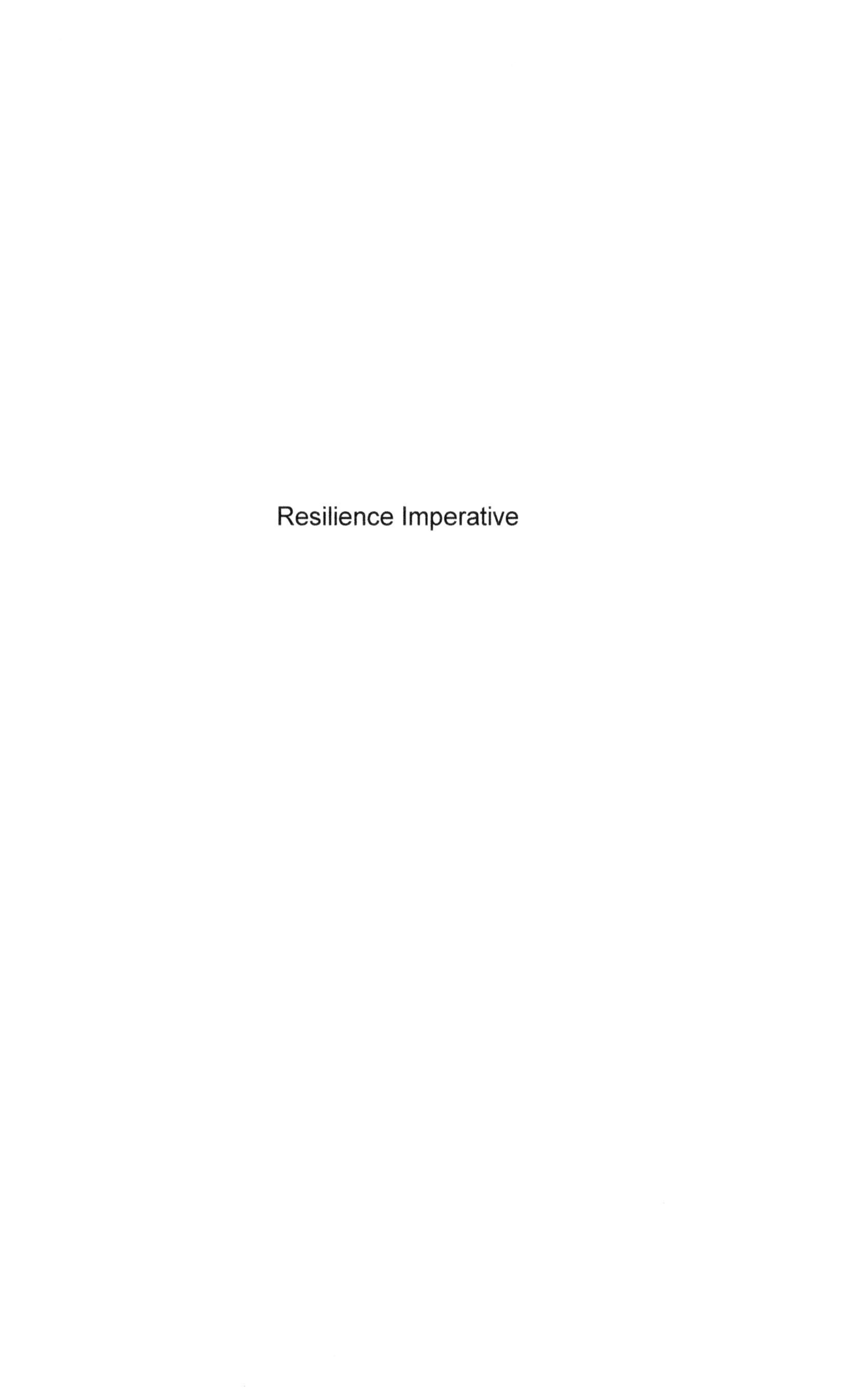

Resilience Imperative

Series Editor
Françoise Gaill

Resilience Imperative

Uncertainty, Risks and Disasters

Edited by

Magali Reghezza-Zitt
Samuel Rufat

First published 2015 in Great Britain and the United States by ISTE Press Ltd and Elsevier Ltd

ISTE Press Ltd
27-37 St George's Road
London SW19 4EU
UK

www.iste.co.uk

Elsevier Ltd
The Boulevard, Langford Lane
Kidlington, Oxford, OX5 1GB
UK

www.elsevier.com

British Library Cataloguing-in-Publication Data
A CIP record for this book is available from the British Library
Library of Congress Cataloging in Publication Data
A catalog record for this book is available from the Library of Congress
ISBN 978-1-78548-051-5

Printed and bound in the UK and US

Contents

Foreword

Increasingly, we tend to speak of "resilience" in the wake of a catastrophe, or when faced with a danger –in a crisis situation or in relation to a distributed event. Indeed, as employed in dynamic and changeable situations, the term "resilience" characterizes both the status of a system and the processes or intrinsic qualities of that system. It is part of dynamics of change and is, therefore, appropriate when speaking of the issues of global change and climate change. Contrary to permanence, resilience goes hand in hand with the dynamics of evolution of a system to deal with a disturbance, destruction and also in the stages of its reconstruction in search of stability.

At the point of intersection between a "before" and an "after", although it is sometimes difficult to define, resilience would act as an indicator of the state of health or the degree of development of a system, whether man-made or otherwise, or an ecosystem in its entirety, of which humans are an integral part, in its ability to bounce back after a change – to be resilient.

From one spatial scale to another, on one timescale or another, and also depending on the levels of organization and commitment of the various actors, or indeed the ways of working of the localities, the same notion of resilience is used both by operators in the field and by academics, in ways specific to each of them.

What kind of resilience do we really mean, though?

This book sheds light on the polysemic nature of the concept of resilience, which is used in psychology, physics and ecology, but has been appropriated by other disciplines, which draw from it but also distort it, to some extent. Through the analyses presented in these pages, the authors demonstrate, in particular, that "by drawing on notions taken from ecology with no discernment or strictness, eclipsing the methodological and theoretical debates over human and social ecology [...] we are rebiologizing and depoliticizing issues which are, fundamentally, social and political ones".

With the aim of providing precision and clarity, the authors begin by offering perspective on the theoretical dimension of resilience, looking at it in relation to other concepts such as "vulnerability", "uncertainty", etc., highlighting strengths and weaknesses, paradoxes and ambivalence, the interpretations and the issues that these terms raise. Using a wide variety of examples chosen for illustrative purposes in varied contexts and situations in the urban environment, in the context of climate change, etc., the authors also show that resilience is the object of power games and varied representations, which manifest themselves, notably, in the discourse of those who use them.

With this book, Magali Reghezza-Zitt, Samuel Rufat and the various contributors help to consolidate the concept of resilience and make it a tool for use by states and societies. They outline and open up a debate which is essential to better help societies face worldwide change, the degradation of biodiversity and climate change, with a view to ensuring a sustainable, liveable future for everyone.

Agathe EUZEN
LATTS, CNRS, Paris

Introduction[1]

"Some cities do better in the face of disaster than others. It is tempting to describe apparent success in terms of resilience and apparent failure in terms of a shopping list of explanatory variables. This is too simple" [COM 10].

The beginning of the 21st Century has been marred by a succession of environmental crises, whether geopolitical, economical or social. This is in the context of a world dominated by uncertainty, where societies themselves are increasingly concerned for their own safety. As a result, resilience seems to have become the answer provided by international organizations and public powers, the alpha and the omega of risk and crisis management. Resilience has become imperative at the global level since the Hyogo summit in Japan and the adoption of the United Nations (UN)'s *Hyogo Framework for Action* (2005–2015). The UN organisms are now meant to help communities in becoming more "resilient" in the face of crises that threaten their development.

Thus, resilience is particularly in vogue: it is a buzzword [COM 10, p. 1], which, like the "buzzing" of insects, makes noise,

Introduction written by Magali REGHEZZA-ZITT and Samuel RUFAT.

1 This book starts with a research seminar held at l'Ecole Normale Supérieure in Paris, France, between 2009 and 2013. The summaries are accessible in French at: http://www.geographie.ens.fr/-Resilience-urbaine-.html. The writing of this book was finished in November 2013.

reflects activity and agitation, but which ultimately tires the ear and results in exasperation and irritation. We try to swat it away, but it comes back all the louder.

In the United States, resilience is everywhere: in the mouths of experts, politicians, journalists, community movements and the general public. In Europe, this Anglo-Saxon influence is increasingly strong, spreading alongside other notions such as sustainable development or governance, in professionals as well as in elected officials. It also receives significant amounts of enthusiasm in the world of science. While it was initially limited to physics, ecology and psychology, resilience has made a spectacular entry into the fields of risk, hazard and disaster management[2]. It is also starting to appear in research on social and spatial processes over varying time periods, linear or non-linear changes in society and territories around issues of transition, durability, dialectics between breaks and continuity, etc.

I.1. Resilience, polysemy, cacophony or quandary?

Resilience refers to the ability to bounce back, recover and rebuild after a shock, a disturbance or a crisis. It is a property which means that, no matter what the events endured, an individual, society or territory does not disappear. It is the process that allows them to deal with disorganization, loss and damage to maintain themselves and endure despite the obstacles emerging from the environment, history or existence. It is also the result of this process that is noticed – sometimes celebrated – and can lead to recovery or even rebirth.

Resilience refers to both the ability to absorb and digest the rolls of the die and the ability to always rise up again from one's ashes. To a certain degree, it makes whatever or whomever that presents it invulnerable, not because it protects from threats or injuries, but because it prevents collapse. It facilitates the overcoming of damage

2 We will return later to the neologism of cindynics introduced by G.Y. Kervern during the founding conference held in the Sorbonne in 1987 in a very global sense, to regroup, out of convenience, all things related to the sciences of danger under one umbrella term. We therefore stray from the precise definition used by Kervern.

and disaster, through a return to a "normal" situation, or even in some cases to a "better situation". When the lessons of a crisis have been learned, resilience is presented as an opportunity to build (or rebuild) something safer, fairer, bigger, more efficient, etc.

Thus, resilience is a highly desirable horizon. It helps us in dealing with contingencies, fulfilling requirements of resistance to dangers and persistence through variability. It also becomes a catalyst of research credit. It is full of a double promise: not only an ideal response to crisis, but also that of public funding. In this way, it opens up several new perspectives at a time where the recurrence of natural, technological, health-related and environmental disasters seems to be marking the end of traditional management policy. Everyone wants a part of it, often with the fervor and vigor of new converts. However, the extrapolation of this concept beyond its fields of origin can be a serious issue.

I.2. Defining resilience

The increase in the number of references to "resilience" has resulted in a large scope of the use of the term, leading to a dilution of its meaning, to the point where the word has started to have contradictory designations. The breadth of definitions can not only be explained by the progressive broadening of the field of applications of resilience, but also by the involvement of a variety of actors, researchers and administrators from different fields. Each offers their own interpretation and definition, depending on their own questions, contracts and protocols. The recent trend feeds polysemy, while disagreement and afferent saturations transform resilience into a cacophony, risking its transformation into either an empty shell or a simple varnish applied to the same old concepts.

For some people, the increasing plurality of designations is the very symptom of its vacuity, and the multiplication of solid definitions and of fragmented approaches are signs of its sterility. Thus, we could consider that it is the hype effect that discredits resilience as a scientific concept, or even as an operational tool, seeing as this trend results in cacophony, overinvestment and the quandary of upmanship of over-bidding.

On the contrary, it is possible to seize this polysemy as an asset, without giving up on the validity and the fertility of the concept of resilience. This requires the consideration of the diversity of approaches, and the cross-sector exploration of resilience, without limiting it to one frame, or reducing it to a single definition, which would only be another to add to those that have already been suggested. All the authors who have contributed to this book do not share the same approach of resilience, and do not all agree on the status of this notion. Some of them actually have very critical views of it, their reservations not only being theoretical and methodological, but also political. In any case, they have provided analyses that have captured the wide range of possible approaches. This is what explains the fact that, depending on the chapters, the definition of resilience varies. However, it is always clearly explained, and compared with other approaches, and the stakes surrounding it are discussed, showing that the different views bridge cross-disciplinary insights.

I.3. Resilience put to the test: the theoretical issues

The advance of resilience on the international stage, and its mobilization by very different stakeholders, results in overinvestment. Clear contradictions between statements, theories and practical solutions lead us to question resilience and to put it to the test. The trending effect is real, but should resilience be rejected as an empty concept with no future as a result?

Resilience generates a number of theoretical problems. Beyond the initial difficulty in defining it, resilience is thought of in reference to a number of connected concepts, which it tends to complete, inflect, or even replace. This is the case for vulnerability, a key concept in the risk and disaster management, with which it shares complex ties. Thus, resilience encompasses questions of social, spatial and temporal scales. Who is resilient to what? Psychology has given emphasis to individual resilience, while other fields look at it more in terms of the resilience of a social group or community resilience. Moreover, resilience regroups the notion of duration and temporality in the question of change. What duration must be considered? Which spatial scale should be used to record resilience aftershock? Resilience requires us to consider the increasing complexity of a world that is more and more interconnected, where each action leads to retroactions on different scales, over vast areas and in uncertain timeframes. This explains both the difficulties of definition and formalization.

This leads us to the question of systems. Resilience is indeed defined in ecology to formalize the question of the timelapse systems need to return to the after a disturbance. For a long time, the term “system” has been strongly linked to the term “resilience”, without a precise definition. It can be used metaphorically, inasmuch as it can express the notion of the complexity of reciprocal interactions, which are found in the majority of objects to which resilience can be applied (social groups, areas, critical infrastructures, etc.), as well as much more rigorously (complex systems theory and modeling and models). Sometimes, the system no longer refers to that which is resilient, but rather to the method used to think of resilience (in this case, we talk of systemic approaches). However, referring to systems implies a certain number of precautions, not to mention that the outlines of the system – or even subsystems – must be charted.

I.4. From practical application to critical examination

The second test is that of application. The elasticity of resilience gives it a practical advantage, as it facilitates its use in very different fields of application. However, even if resilience is presented as a promising answer to crises and uncertainty, the passage from theory to practice remains challenging.

Everyone henceforth wants new solutions to create, confront and improve the resilience of social and/or spatial systems. Resilience refers more here to a capacity, developed upstream of the disturbance, than to a recovery process. It relies on the convergence of different organizational, technical, social and cultural factors. The efforts mainly focus on the anticipation of disturbances, mitigation and learning of emergency management devices, and also on the available resources for the post-crisis period.

In this context, is resilience a turning point, a new framework for action, a new paradigm in disaster management, or must it rather be considered as a mere change of focal length?

In practice, the strategies that aim at producing resilient systems are quite varied. A first approach focuses on the material dimension of resilience. Following material and functional vulnerability reduction, it involves adapting the components of systems or rethinking their localization, so as to make them less vulnerable, or so that they absorb shocks more easily. A second field of action has more to do with the modalities of the functioning of the system, and defines resilience as the ability to maintain activity and return to an equilibrium. A final lever concerns the organizational dimensions of crisis management and considers resilience as the ability to deal with the disruption. It refers to the confidence that agent have in their ability to master a crisis. It relies on learning-based approches. It aims to reinforce individual or collective ability in prevention, planning, informing and adapting to cope with crises.

Therefore, we can see that resilience can be found either in anticipation, or in reaction to disturbance and crisis. Proactive and/or reactive, resilience is split into multiple actions and projects, with varying amounts of success, which we will attempt to illustrate in this book through different case studies (adaptation to climat change, reconstruction of New Orleans, redevelopment of East London, etc.). There still exists a debate over the possibility of transforming a retrospective observation (such and such individual, society or region has been resilient) into a prospective tool that would make resilience a framework for action.

However, the application of resilience pushes it into the field of discourse. There is often a gap between discourses held and actual practices. More globally, something may be resilient only because it was decreed at some point by someone. The quality of one stating resilience, the choice of the moment, the place, the manner in which the sentence is pronounced, or even set up, are not neutral factors.

Moreover, the type of resilience depends on the agents, their positions, their intentions and their decisions. In a way, resilience can even end up saying more about those who talk about it than about the facts. Resilience further allows to investigate the question of memories and the link with the past. Resilience must also be considered as a discursive construct: the application of resilience, whether as the injunction of large international programs or more local development projects, questions the limits and the contradictions of some political and operational uses. In fact, it is mandatory to first remove any underlying ideological or moral assumptions so as to put resilience narratives and discourses into perspective.

I.5. Bibliography

[COM 10] COMFORT L.K., BOIN A., DEMCHAK C.C., *Designing Resilience. Preparing for Extreme Events*, p. 272, 2010.

1

Defining Resilience: When the Concept Resists

Undoubtedly, the success of resilience arises not only from the positive connotations it evokes, but also from the inherent equivocity of its uses [FOL 02, DOV 92]. Being an abstract term, it is rarely used in common speech, and has therefore little intuitive meaning. Resilience provides a flexible outline and undeniable plasticity. For this reason, the term has many definitions, each providing different contents, depending on needs and practices. It also includes a "rainbow of meanings" to use D. Collin's [COL 10] expression – these resonate together as they become apparent, but also sometimes contradict one another.

This polysemy is both an advantage and a disadvantage. It is an advantage as plasticity allows us to take resilience and adapt it to the framework of the relevant discipline. It can be introduced in an ever-increasing number of fields, and open up a large number of research paths. It is also a disadvantage as the large number of definitions means that the concept is vague, decreasing its appeal from a scientific and practical point of view: the appearance of an imprecise portmanteau or catch-all quickly becomes repulsive.

Chapter written by Magali REGHEZZA-ZITT, Serge LHOMME and Damienne PROVITOLO.

The goal of this chapter is not to add yet another definition to the list, but rather to establish an overall account, a sort of map of current uses: who is saying what, in which disciplines and why? What paradigms are implied by each definition? Why do some definitions end up contradicting each other, at least superficially?

1.1. A multidisciplinary construct

Before becoming a fashionable concept in cindynics, resilience had been used in three fields: physics, ecology and psychology. Each of these fields formalized the term, respectively, contributing to its polysemy.

Etymologically, the term "resilience" derives from the Latin "*resilio, resilire*", which refers to the action of jumping backward or bouncing back. In French, *resilire* has given rise to the word *résilier* (for a contract, for example), meaning to cancel, or break. "Resilience", however, is a word which is not commonly used in everyday speech. However, the adjective *resilient* had existed in English since the mid-17th Century, meaning to "bounce back" [KLE 03]. It is this meaning that comes through in the noun "resilience" (or resiliency).

The word "resilience", as understood in its etymological sense, is therefore very recent in modern-day French: unlike for the English-speaking world, it is first a scientific concept before being a vernacular concept.

1.1.1. *Material physics: resilience and resistance*

The first scientific use of the terms "resilience" and "resistance" was in relation to the material physics. The term "resilience" was used in works by Charpy in 1901, to measure resistance at the point of failure of a material. The resilience test involves breaking a notched piece of steel, called the Charpy V-Notch, using a pendulum impact-testing machine. In this definition, resilience is opposed to brittleness, since the more resilient a material is, the more is the amount of energy required for the material to break.

That is why there is some confusion between resilience and impact-resistance. However, the research that followed those by Charpy do shed light on this by introducing the notion of ductility. Ductility is also opposed to brittleness, and also to the ability of the material to deform without breaking. Different materials are therefore more or less ductile depending on their elasticity (the ability of a material to return to its initial state following an impact) and plasticity (the material does not return to its initial state following an impact or deformation). Resilience is then dependent on the elastic and plastic properties of the material and no longer on the property of resistance.

1.1.2. *Psychology: resilience, processes and properties of overcoming, reconstruction and reinforcement*

Psychology was the first discipline to use “resilience” as a disciplinary “construct”. The first use of the term appeared in the 1940s. John Bowlby first used the word in reference to “*les personnes qui ne se laissent pas abattre*”, which can be understood as *people who do not let themselves get knocked down* [BOW 69]. It was then used in studies on children who were expected to develop mental pathologies as a result of emotional trauma [WER 89]. The concept was then distributed in France due to the ethologist, psychiatrist and psychoanalyst Boris Cyrulnik [CYR 01].

Resilience in psychology is therefore seen as an explanatory model that is applied to traumatic events or traumatogenic contexts (for example, living in very unfavorable familial and social contexts). The term describes and refers to the functioning of individuals or groups (e.g. families), which manage to put up with and overcome the destructive consequences of traumatizing situations. They live with the traumatism, maintain a certain quality of life and take the least amount of damage from the traumatic episode.

Resilience also seems to refer to both the resources and the processes that allow us to face the deleterious effects of a traumatic experience, and to rebuild following it. It is not, however, limited to the processes of recovery that follow the traumatic situation [CYR 06, ANA 09]. It does not entail a return to a state that preceded the traumatism: the subject adopts *another* type of development *following* the trauma, due to the abilities of adaptation, and, as a result, is strengthened.

Two schools of thought do disagree, however, in explaining this adaptation. The first school highlights individual competences. The second school stresses the role of factors that are external to the individual, for example their environment. Whichever the approach taken, resilience has nothing to do with the notion of invulnerability; it corresponds to the ability to return to a "normal" lifestyle, despite the injury, without becoming fixated on this injury. Far from a miraculous or magical form of scarring, resilience is therefore neither a vaccine against victimization nor an anesthetic against suffering.

1.1.3. *Ecology: resilience, persistence and adaptation*

Resilience also started to be used in ecology, but this was not without some debate, and was not done in a linear fashion [PIM 84]. The founding text is that of C.S. Holling, published in 1973, which defined resilience as a "measure of the persistence of systems and of their ability to absorb change and disturbance and still maintain the same relationships between populations or state variables" [HOL 73].

The author opposes himself to theoretical models that were dominant at that time, as they did little to consider random phenomena to which ecological systems are submitted. Indeed, ecosystems appear to be complex nonlinear systems. Holling suggests relying on more qualitative approaches, less centered on a return to the equilibrium, so as to focus on the study of the persistence of these systems. Indeed, according to him, an uncertain and unpredictable world requires a consideration of the ability of the system to absorb and "accommodate" unexpected events in the future. It is therefore necessary to analyze the "strategies" that explain the ability of ecosystems to ensure their own survival in difficult conditions.

We are therefore no longer talking about a return to equilibrium, which corresponds to a return to the initial state that characterizes the stability of the system. For Holling, this return would actually oppose resilience: persistence and stability would become contradictory, and resilience and persistence would become synonyms. Progressively, he reviews his own definition and adds detail to it: a system that persists despite a disturbance by changing its qualitative structure cannot be considered to be resilient [HOL 01]. On the contrary, a resilient system must maintain a certain qualitative structure, so as to continue to function, or fulfill part of its functions. In the opposite case, we can assume that bifurcation has taken place. As a result:

> *Resilience corresponds to the ability of a system to absorb disturbances, or to the maximum amplitude of a disturbance that can be absorbed by a system before it changes its structure by changing the variables and the processes that control its behavior* [HOL 01].

Starting with these pioneering works, two distinct schools have come into opposition: "ecological resilience", which arises directly from the work of C.S. Holling, and "engineering resilience", which derives from a more traditional view of stability. Thus, for "resilience engineering", a resilient system is a stable system that is permanently close to a state of equilibrium [PIM 84]. Resilience then refers to the ability of structure undergoing a sudden shock or continuous pressure to resist without transformation. The measurement is made from the resistance and the speed of the return to equilibrium. However, for "ecological resilience", a resilient system maintains its essential functions and structures, not by maintaining a single state of equilibrium but by going through different states of equilibrium (stable and unstable).

1.2. Transfers in cindynics

There is a notable amount of interest in resilience today; it is omnipresent in cindynics. Large international agencies have even made it a main focus: it is both the objective and the framework of new management policies[1].

The rapid diffusion of the word is all the more surprising as it affects academic spheres just as much as it does operational spheres. Resilience can be interpreted as a new moment in the history of risk and hazard management and the understanding of disasters. Indeed, although the concept has been used for many years, it only became central around the beginning of the 2000s.

1.2.1. *The early 2000s*

The recent fascination around resilience can be explained by the conjunction of a triple context: the recurrence of deadly disasters that were thought to be under control (at least in developed countries), with hurricane Katrina playing a major role in this regard; the September 11 attacks in 2001 in the United States and the focus on mass terrorism; climate change and the questions that arise from it [QUE 13].

Chronologically, resilience seems to have been first used in work on climate change [CUT 08]. Beyond the stormy (and very media-present) debates on the role of mankind in global warming, the question of climate change was first progressively tackled during the 1990s through analyses of cindynics. First, climate change would appear to likely affect the intensity, frequency and even distribution of the climatic and water-related incidents, particularly cyclones, draughts and floods. Therefore, it becomes a possible factor in the creation, worsening and increasing of disasters. As a result – and this is the second point – it can be indicated as a meta-hazard that threatens vulnerable populations that are exposed to it, or more precisely, that are exposed to the threats that it induces (not only

1 For example, in 2005, the UN adopted, as part of the International Strategy for Disaster Reduction (ISDR), the Hyogo framework (HFA), a 10 year plan (2005–2015), to "make the world safer in the face of natural disasters". This plan, adopted by 168 countries, has the following sub-title: "building the resilience of nations and communities to disasters".

natural threats, but also sanitary, economic, geopolitical and social threats).

The main problem is that we do not know the nature or the intensity of the effects of climate change, nor do we know the real extent of the exposure to the threat (how many people, where and how?), or the speed at which the changes can operate [DAU 13]. As a result, the classical response of cindynics – which is based on the impacts caused by incidents on social and territorial systems and opposes them with technical attenuation and protection solutions[2] – is not fully adequate [DAU 07]. Those in charge then promote the notion of adaptation to change, which no longer implies a simple physical *reaction* to the impact, but rather a *dynamic* social response: when it is not possible to act on the physical processes, it becomes necessary to develop the ability of populations to adapt to the new climactic situation and the uncertainty that it brings [GAL 06]. B. Quenault talks here of:

> *A radical change of posture which would mark a shift from a technicist paradigm toward an eco-systemic one, where it is no longer about 'fighting against' the evolution and changes taking place, but rather 'making do' with them* [BER 10, QUE 13].

Experts also recommend reducing greenhouse gas emissions, limiting deforestation or decreasing the effects of heat in towns by increasing the amount of vegetation. Other forms of adaptation exist, such as the climatic migration – temporary or permanent – of populations, or other forms of societal adaptation, which can be more or less punctual, allowing groups to live alongside the danger, or to integrate it in the functioning of the in societies and territories.

2 The term used in cindynics is *mitigation*. This refers to the actions aiming to stop the incident from taking place, as well as those that attenuate the amplitude, the intensity or the frequency, and, by extension, the "structural" measures taken, which aim to reduce the exposure of populations by erecting barriers (dams, snow fences) or strengthen the physical resistance of technical protective infrastructures.

In this context, climate change is seen not only as a creator of disturbances (incidents caused, or incidents made worse by climate change, particularly natural changes) but also as a "meta"-disturbance, which calls for a response from the affected system. This goes back to the notion of resilience as developed in ecological resilience. Starting in 1981, Timmerman talked of the "resilience of societies" to climate change [TIM 81] and operated the transfer of ecological systems to social systems. Later, the notion of complex socioecological systems, developed notably by the *Resilience Alliance,* a multidisciplinary research group that studies the dynamics of adaptive complex systems, helped in combining social and ecological aspects, with towns and cities proving here to be a privileged area for the application of these approaches. We can note that resilience to change and resilience to disturbances, while strongly linked, do not fit into the same exact temporality: change does imply a long, nonlinear process, with variable rhythms, while a disturbance is first seen as a punctual impact, a rupture causing a degree of unbalance [TOU 12]. Resilience to climate change therefore helps stress in having the idea of the *continuous* adaptation of sociospatial systems, which "retrospectively react continuously to the changes in their environment that they themselves participate in creating" [QUE 13].

Resilience experienced an increase in popularity following the attacks of September 11, in 2001. Following the attacks, political and ideological discourse, reflecting the founding myths of the American nation, highlighted the importance of resilience. This attitude reappears in the United States after each important catastrophe. It was analyzed, with regard to the large fire that affected Chicago on October 8, 1871, by Harter [HAR 04], or in the collective work by Vale and Campanella called *Resilient Cities* [VAL 05]. It magnifies those adaptive capacities inherited by the pioneers: Americans are able not only to face disasters, but also to rebuild big and better afterward so that the catastrophe may be overcome collectively and individually, resulting in a positive outcome. This way of telling the story of the catastrophe, mixing patriotism with hints of religion, was followed, in the case of 9/11, by a second voice, which highlighted the extraordinary weakness of the country, despite it having then reached a level of power never seen before in its history, and indeed in the history of the world. This paradoxical vulnerability of the United States, faced with threats as unpredictable as they are unconventional, requires a response that cannot be found in the traditional foundations of security. As a result, political and academic discourses converge around resilience: it refers to this ability cope with threats whose occurrence is inescapable, that, while unpredictable and unpreventable, can be overcome due to an ability to adapt that confronts adversity, overcomes crises and comes out of it even stronger. Here we come across again in part the concept as developed in psychology.

Another major event was the destruction caused by hurricane Katrina in the south of the United States in August 2005. The damage was enormous [HER 10]. New Orleans became a life-size laboratory of the process of *recovery* [HER 09]: in it, the mechanisms of post-crisis management could be observed, as could material rebuilding, economic and demographic recovery, the symbolic rebirth of urban life, the remaking of the city's identity and that of its inhabitants. Faced with political discourse that called for using the disaster of hurricane Katrina as an opportunity to rebuild a better, safer, more dynamic and fairer city, researchers were able to deconstruct this rebuilding process [HER 10]. They looked into the reality of resilience and added new questions to it. The lessons learned from Katrina meant that it became possible to distribute the concept of resilience – which was greatly used *in situ* – to professionals on a large, international scale.

1.2.2. *A response to theoretical and methodological issues*

Another way of thinking of the success of the concept of resilience is to consider a response to a theoretical, methodological problem [COM 10a]. Particularly, it can be viewed as a solution that enriches – or even renews – the devices used in management and politics that contribute *in fine* to risk reduction [DAU 07].

The disasters that occurred in developed countries from the start of the 1990s presented a whole new dimension. First, we can observe a series of technological failures, and also of intentional attacks (power cuts in Canada or in New York, attacks in public transport in London and Madrid, etc.), whose effects were disproportional to the disturbances that caused them. Localized incidents cause damage that spreads quickly and paralyzes everyday life as well as the economy. These chain reactions are difficult to predict and anticipate, and traditional management plans often prove to be inadequate in these situations. At the same time, well-known problems that have been around for a long time produce unexpected effects: in 2010, the eruption of a volcano in Iceland paralyzed air traffic; in 2011, a magnitude 9 earthquake in Japan, causing mainly low amounts of material damage, resulted in a tsunami that in turn destroyed the Tohoku coast, resulting in thousands of victims. The water reached the nuclear reactor of Fukushima Daishi and resulted in a major nuclear incident, the worst since Chernobyl; in 2012, hurricane Sandy paralyzed Manhattan and destroyed part of the New York subway.

The disasters of the 21st Century have changed in nature: new types of threats now appear, called "new risks" [GOD 02] in the French literature, that are analyzed using the systemic approach [LAG 93]. These risks involve complex damage dynamics [REG 06]. They are ubiquitous and multiscalar. The effects become distributed very rapidly beyond the initial point of impact, and the damage shifts both temporally and spatially [DAU 13, SER 13]. By producing new spatial configurations, particularly reticular organizations, globalization can be seen as the cause of quick diffusion phenomena, while the complexity of contemporary territorial systems, notably urban complexity, results in multiple interactions between incidents and weaknesses [GOD 03]. Domino effect reactions occur increasingly, and uncertainty grows.

However, and this is the second observation, the understanding of the change in the nature of the threats has been followed by a distancing of States, in the name of liberalization and of the economic crisis, meaning that managers are now faced with dealing with new risks, with reduced means, despite the fact that some of the old threats have not yet been eradicated. Comfort *et al.* summarize what they call the "rise of resilience":

> *Dominant trends such as globalization, increasing interdependence and complexity, the spread of dangerous technologies, new forms of terrorism, and climate change create new and unimaginable threats to modern society (...). The concept of resilience holds the promise of an answer* [COM 10b].

In this context, resilience provides an opportunity that is threefold. First, it allows for a more positive, more optimistic outlook, unlike the negative vision spread by the notion of a "vulnerable society" [FAB 87]. It creates a more consensual project, which is more integrative, a horizon of expectation and of action [LAL 08]. Resilience also opens up the possibility of moving the objectives of management, and therefore reaffirms the legitimacy of the managers: while it may not be possible to stop the crisis from happening, it is still possible to act so as to prevent the catastrophe [REG 12].

As a result, two strategies come into opposition [WIL 98]. The first strategy, based on the principle of anticipation, could provide a rapid response to known problems by establishing rules and procedures (but this strategy has some problems dealing with unpredicted and nonanticipated events); the second strategy, based on resilience, leads the way for dealing with uncertain and unanticipated events[3] [DAU 13]. The first strategy is often called the strategy of attenuation (or mitigation) and involves deploying a whole range of legal and administrative techniques to overcome the risk and to stop or limit the crisis. This type of strategy relies on the ideal of control associated with modernity, with an *a priori* definition of actions on the causes with the goal of controlling the effects. The second strategy is called an adaptation strategy, and is often opposed to the previous one by way of the metaphor of the oak and reed.

3 In the first case, the vulnerability of the system is comparable to a state that is a function of the resistance to disturbances, whereas in the second case, it is a process that is built and deconstructed in the daily establishment of actions for the management or prevention of crises and disasters. These actions can be read as much at an individual level as at a collective or institutional one (see Chapter 2).

Finally, by highlighting the autonomy of individuals and abilities of self-organization, resilience facilitates a circumventing, in part, of the issue of the reduction of the means allocated toward the prevention of catastrophes: it shifts the responsibility of management toward the populations, forming here a pragmatic response to state disengagement.

1.3. Defining resilience

The construction of the concept of resilience largely explains its polysemy. Each field of science has taken the term and adapted it to its own positions, both theoretical and methodological. Moreover, some authors note that:

> *Definitions of resilience are less precise in the human and social sciences than in the "hard" sciences as these proceed mostly by analogy and mostly depend on the epistemological reference model and the context in which the notion is being studied* [QUE 13].

In any case, resilience is defined in different ways depending on the author and the field. However, they all have the same starting point, which reflects a shared intuition: when a system is disturbed, or suffers from an impact, either it recovers, or it disappears. To describe this simple trajectory, we use the term "resilience". The problem is that the trajectory in question is anything but linear, and to show this concept of resilience, it is so necessary to bring up connected notions, which can differ and are not always compatible.

1.3.1. *The "rainbow" of meanings*

When a hazard, which threatens a society, a territory or a technical system, becomes a reality, the shock that results leads to material damage and financial and human losses, which can be associated with more or less severe dysfunctions within the system considered. Above a certain threshold level of disorganization, a crisis and a disaster is considered to have ensued. These are states in which the function of the system is disturbed in a manner meaning that an extraordinary and abnormal situation is being encountered (the normal situation being the reference situation, most often assimilated with the initial state). Thus, several authors [QUA 05] have suggested criteria to distinguish disasters from crises and disasters, but without clearly stating the threshold level of effects, or the combinations of variables, at which the transition from one state to another would take place.

However, most of the time, there is some recovery, rebuilding, restoration, renewal and a return to equilibrium or to normality etc. These situations associated with the notion of resilience and its etymology all go back to the idea of bouncing back.

Resilience is therefore placed at the center of a cloud of interconnected notions (Figure 1.1), which in reality describe different situations, linked together but not similar, and sometimes even opposites [LHO 12]. Thus, while restoration implies a return to the previous state, which can be more or less effective, renewal tends to result in a different state, which can indeed draw inspiration from and rely on its heritage, but which stresses the importance of novelty.

Figure 1.1. *The cloud of meanings © Serge Lhomme*

More generally, these notions bridge an important ambiguity.

– for some authors, resilience can be a result, a process, or both [CUT 08]. As a result, resilience expresses the fact that the system is able to rebound and deal with the threat. As a process, resilience refers to the succession of consecutive responses to the disturbance and goes back to the mechanisms that lead to the renewal of the system and to the appearance of new trajectories;

– for others, however, resilience is a property, an intrinsic quality of a system, an ability that shows itself at the moment of the impact, but which is already present before this moment.

When resilience is a process, it gives rise to the state of resilience. In this context, to say that a system or a stake is resilient, involves stating, *a posteriori*, that it has succeeded in maintaining itself despite an impact and managed to overcome the crisis that resulted. Resilience then functions as an interpretive model, whose value is primordially heuristic. This is a diachronic perspective: over long time periods, we consider resilience to be a dynamic process, present in the long run, with its own temporality and rhythms. The question here is whether this basis for resilience does exist, or whether a system is called resilient simply because we interpret certain processes according to this model, and this is all the more true as the concept is less precisely defined and more malleable [VAL 05].

When resilience is an intrinsic quality, referred to as an ability, capacity or capability [SEN 85], the analysis does not focus on the result, or on the impact, but on the system exposed to the impact. The relationship with time is different: resilience exists before the impact– it is a potential, made apparent by this disturbance. It is therefore a-chronic in a certain manner. Moreover, if resilience is a quality, it can be innate or acquired, but, as opposed to the previous definition, it does not depend on the *a posteriori* observation: we can be resilient without knowing it. The capacity for resilience can also be explained by several factors (biophysical, social and spatial) and it is possible, once these are identified, to carry out a prospective – and therefore operational – approach to improve the potential for resilience.

There also exist interpretations that are divergent within each of these approaches. If resilience is a process or a state/result that relates to the observation, it depends on the view of a third-party, which implies establishing criteria that will facilitate for the determination of whether a system is resilient or not. This leads to the issue of the nature and the thresholds of the qualitative changes that can establish whether resilience is present or not. In a very schematic manner, an impact can lead to the definitive disappearance of the system, or the maintenance of the system in a state more or less close to the initial state, or even the bifurcation of the system, which refers to a change in its qualitative structure. However, there is no consensus on the relation between the degree of transformation and the state of resilience. For example, for some, there is no radical opposition between bifurcation and resilience. For others, however, resilience means stability, and refers to maintenance without any change. Others still understand this stability as meaning a differential adaptation of the components of the system: some elements can change, but the core, whose perimeter needs to be defined, is unchanging. In this view, it is possible to have a system that rebuilds itself, but which is not resilient [PRO 09].

The definition that makes resilience a property is just as ambiguous. It is often described as an adaptive capacity, either reactive or proactive, and can be split into different ways:

– resilience as the ability to withstand or resist an impact;

– resilience as the capacity to absorb an impact, which implies relative plasticity;

– resilience as the capacity to react to the impact, which can go as far as the capacity for self-organization;

– resilience as the ability to recover or rebuild, using internal and/or external resources;

– resilience as a capacity for anticipation, which favors previous capacities, a capacity for learning, which allows for lessons to be learned from the event and therefore for reinforcement to take place;

– resilience as the system's ability to maintain its integrity and to return, or *bounce back*, to a state that can be the previous state (restoration), the state of equilibrium, the normal state (which is hard to define), when it suffers from a disturbance.

These definitions are not always compatible and are the subject of energetic debate. For example, the contradiction between resilience and resistance can be underlined: these are synonyms for those which consider that resistance gives the system time to adapt, but opposites for those for whom a capacity of adaptation implies flexibility and plasticity (while resistance involves opposition and rigidity) [ADG 00].

1.3.2. *Recognizing resilience*

Faced with the multiplicity of the definitions, two problems arise: how to define a resilient system, or, by default, how to define a nonresilient system? The response to these two questions, as simple as they may be, is far from obvious. Even if we assume there exists an interpretive model which is able to subsume a multitude of varied but recurrent situations, under the encompassing word "resilience", empirical observation shows that there is often a gap between this and the diagram of disturbance/destruction/reconstruction.

In this way, is a system that avoids a crisis, because it is ready, really resilient? When a society is prepared to deal with a threat, permanently adapting to it, the risk of a disaster ends up being but an accident of daily life. Thus, urban fires now only cause limited, localized damage, while until recently they could destroy entire cities; some sanitary risks, which in another time were deadly, have been close to eradicated through effective vaccination campaigns. For some, the absence of crisis is an indicator of strong resilience, as it implies that the system is able to effectively absorb impacts so that they are hardly even noticed, but for others, who think that resilience starts with the disturbance, this is contradictory as resilience can only exist if there is a damaging event great enough to destabilize the system at least for some measure of time. There is some opposition here between resilience as a return to equilibrium over a relatively short time period, and resilience as the progressive adaptation of a system over a long time period.

Inversely, is it possible to talk of resilience when considering the case of a system that would avoid a crisis not because it is ready for it, but because the threshold above which the system would not have been able to resist has not been reached? Augendre reminds us as an example in Japan "adaptation to risk happens daily for those living near volcanoes whose activity is moderate but sub-permanent, and allows them to accommodate adequately" [AUG 05]. However, the Kobe earthquake of 1995, or the tsunami that devastated the coast of Tohuku in 2011, show that above a certain damage threshold, the society's capacity for adaptation is overcome, at least temporarily. To what extent can we then speak of the resilience of Japanese society?

Another problematic case was the terrible earthquake that hit Port-au-Prince on the 12th of January, 2010. When a system is undergoing a latent crisis and is in a situation of deep and durable decline, the disturbance usually only highlights existing trajectories: what does "resilience" mean at this point? And what does resilience mean when the crisis becomes the "normal" state?

If we look at the problem the other way around, it is not easier to define a nonresilient system. Intuitively, it would be a system that suffers from a crisis and then disappears as a result. However, this case is very rare, if not exceptional: for example, history only provides a few examples of the definitive destruction of socioterritorial systems [VAL 05]. The example of towns is particularly enlightening. In his work on what he refers to as "nomadic towns" [MUS 02], Musset showed how the site, name, etc., of towns destroyed by natural disasters persist. Archeology provides several mentions of the "recycling" of sites in the long term, with different functions, uses and spatiality [ROB 13, ARC 98]. Pompeii is an exemplary case, with a tourist renewal of the site in the 19th Century, while a town with the same name has been rebuilt near it. Is this resilience, and if so in what sense?

In the same manner, is a system which undergoes a crisis, and which comes out of it, but radically changed, still resilient? French geographers have written a lot on this subject. Some works, such as those by Aschan-Leygonie on the spatial system of Comtat-Venaissin [ASC 00] or those by Pierdet on Phnom Penh [PIE 08a, PIE 08b], explicitly used the resilience/bifurcation association. Djament preferred the concept of "spatial reproduction" to insist on the interactions between the (re)production of that which is the same, and the production of that which is different, and the stakes involved in the social reproduction associated with urban renewal [DJA 05]. These questions have also been involved as a part of a more long-term perspective, with, for example, the concept of "*pérennité urbaine*" (a term created by French geographers, with no equivalent in English) [VAL 09], which goes beyond the perception of time segmented by ruptures (the disturbances) to go back to an urban and/or symbolic or cultural continuity [BER 08]. Such an approach allows for distances to be taken along the lines of an exceptionalist view of the disaster, and to relocate the crises in relation to "normal" urban fabric.

1.4. Two concepts for a single word

Resilience leads to contradictory definitions, and this wealth can be a source of confusion. The concept then becomes inapplicable, with resilience seemingly referring to a form of unattainable utopian discourse. As a result, it is necessary to demonstrate the coherence between different meanings and uses.

We propose here a hypothesis. The ambivalence of resilience reflects a division into two distinct paradigms, which are complementary but irreducible one to the other. These two trends, which have opposed each other in the past, now constantly exchange and adapt notions, models and tools that they borrow from each other. They share the same vocabulary, and as a result give the impression of conflation. However, they both rely on certain approaches, interpretations, different understandings of risks and disasters. In this way, they provide two distinct views on a same object so that there is not one, but two concepts contained in the single word "resilience".

1.4.1. *Resilience in the "techno-centric" paradigm*

The first trend, *risk analysis*, which to this day still plays a main role, emerged at the end of the 18th Century in Western physical science and engineering, around the idea that it is possible to rationally manage disasters due to the progress of science and techniques (giving rise to the term "techno-centric paradigm"). These techniques must allow for a gradual eradication of damageable physical processes. This paradigm gained structure throughout the 19th and the 20th Century and is characterized by three elements:

– a focus on material damage, which must be prevented at all costs. Incidents are considered as exogenous threats, which come to disturb the equilibrium. It is an enemy that must be opposed by barriers, and if this fails, which must be removed from the land;

– an approach of danger, based on quantitative measurement, which reintroduces statistical regularity where a mere mortal sees only chance: the risk is defined here as measured uncertainty, a definition that was formalized in 1921 by the economist Chicago F. Knight (giving the name “risk analysis” [KNI 06]);

– the use of technical solutions, along the lines of a physical process that must be contained – or even eradicated – from society, which is reduced to a passive sociotechnical system.

As the sciences and techniques progress, means – first reserved for the management of the crisis and for rescue – are moved upstream of the crisis, and more or less grouped together in the category called “prevention”. For a long time centered on the reduction of physical process (mitigation) and on the physical protection of people and of goods, the devices are enriched with measures to reinforce the resistance of constructions (building norms) and with regulatory zoning to control exposure. If preventing material damage is not possible, attempts are made to keep it to a minimum. This damage is seen as a physical fault (construction), a technical fault (breakdown/outage) or an organizational fault, which must imperatively be anticipated. For this, plans must be made that contain the rules to be followed, as much as for construction as for behavior, following a *top-down* logic. In the case of damage, rebuilding must – ideally – lead to the sociotechnical system being safer.

In this context, several interesting definitions of resilience arise:

– resilience as the ability to *bounce back*, in the sense of going back to the state before the crisis, which would involve reacting to the physical impact and overcoming the material damage;

– resilience as the ability to *resist* the impact physically (the opposite of fragility) and therefore to *absorb* it (definition of resilience in physics);

– resilience as the ability to *maintain* functions despite the disturbances;

– resilience as the process of material *reconstruction, supporting the return to activity* of the functions of the system.

These definitions go back to several properties of threatened elements: the physical ability of resistance of the material, the adaptability of structures that enable the continuity of activity, the capacity for anticipation and reaction of organizations. These properties all involve the notion of adaptation, taken here as the result of an exogenous action applied to the system: the technique helps *adapt* the sociotechnical system to the occurrence of a disturbance, reinforcing it to allow it to maintain itself and last despite the impact. Through this logic, the *stability* of the system is highlighted.

Resilience allows for a definitive unification of all of the approaches of the dominant paradigm, without destroying its foundations. It links the three stages of the disaster (upstream – risk situation – emergency/crisis, after-crisis/rebuilding), with each moment interacting with the occurrence of the two others. It associates biophysical and social, technical and organizational, material and functional aspects of risks and of crises. As a result, it enables the adopting of an integrated transverse approach to the term chapters of risk and disaster management.

1.4.2. *Resilience in the social sciences*

The second trend is more closely linked to the social sciences. They fit into the field of natural catastrophes as part of a larger reflection on the nature of interactions between nature and human societies, and have given rise to an alternative paradigm that is sometimes called hazard research (as opposed to risk [REG 06]). Without denying the importance of technical and engineering solutions, these works focus on endogenous factors that are at the origin of disaster and deconstruct their "neutrality" [WIS 76]. By relying on qualitative approaches, they interpret hazards as a social construction: a discursive and cognitive construct, as all hazards are perceived and named; a societal construct in the sense that hazards are produced by societies, over a certain time period and in a given space.

Moreover, the social sciences have certainly taken time to denounce the hegemony (and the ineffectiveness) of the dominant views, said to be “technicist” and “technocratic”. This criticism was all the more virulent as some researchers would inscribe their analyses into some radical perspectives: the fundamental questioning of a capitalist system judged to be unequal, the anti-democratic nature of societies or of international organisms [HEW 83]. This criticism has contributed to provide a rather simplistic view of the opposition between the two paradigms.

While both paradigms are not watertight, resilience has not proved to be a new moment in management, unlike what happened with *risk analysis*. Resilience was already present in these works, either implicitly when the researchers were formalizing adaptation as soon as the 1920s–1930s or explicitly in the 1960s–1970s with the ability to cope with the capacity of adaptation, capacity of autonomy, capacity of resistance and capacity of resilience [BUR 78, BLA 94]. Resilience was then only a component of social vulnerability, a logical consequence of the capacity for adaptation borrowed from ecology and adapted to the singularity of social systems.

For the dominant paradigm, resilience is above all a state, a result that can be modified by technical and organizational potential. However, social approaches believe that resilience is first a process, a construct that fits into the straight line set by previous works on risk, hazard and vulnerability. They highlight the proactive (and not only reactive) character of adaptation by favoring the inherent abilities of individuals in dealing with experiences (and not those of technical and organizational systems), self-organizing and learning lessons from these experiences. They prefer the term “recover” to “rebuilding”, “restoration” or “rebound”. There is no real equivalent of the word “recovery” in French: the term refers to a more or less conscious strategy, one that is more or less active and that goes further than simply material or functional reconstruction. The process of recovery allows us to get back up, to bandage our wounds, to overcome the immediate impact, in time to maintain our form or identity by integrating the change. It cannot be reduced to the physical dimension of resilience. If we use the urban example, it involves as much the *urbs*, meaning the town in all its materiality and functions, as the *civitas*, meaning the solidarity of the citizens and the symbolic dimension of the urban system [HER 10]. Recovery also excludes a return to the identical, if only because a catastrophe leaves a trace in the collective and individual memory.

In this perspective, societies and territories are considered to be unstable systems. They do not stop transforming, and change is inherent to their nature: it is not negative in itself, and does not need to be stopped or favored. While the dominant paradigm focuses on the rupture that is caused by the disturbance event, favors shorter timescales and works with a crisis that is limited both in time and space, social approaches use a historicizing approach. For them, the disturbance must be inscribed in the diachronic thickness of social and territorial systems. While the event can certainly be considered as a rupture that is often brutal and sometimes traumatizing at a certain level, it is also explained by the temporal continuity that gives it meaning. Therefore, the approach favors the regularity of the structure over the instantaneousness of the event in the restrictive sense of something new [LAB 09].

1.5. Conclusion

Two forms of resilience begin to appear: a technical and organizational resilience, first reactive and then proactive, but ultimately operational and prospective; a socioterritorial resilience, an interpretation model that is above all heuristic. The first form of resilence enables the renewal and reinforcement of the dominant paradigm of *risk analysis*, whereas the second form leads the way for the continuing and furthering of analyses in *hazard research* developed over the last century.

Clearly, the two definitions are not impermeable to each other: they enrich each other mutually and are constantly exchanging. However, they correspond to practices and paradigms that are fundamentally different. They can cause reciprocal incomprehension as, although the same word, they do not refer to the same thing.

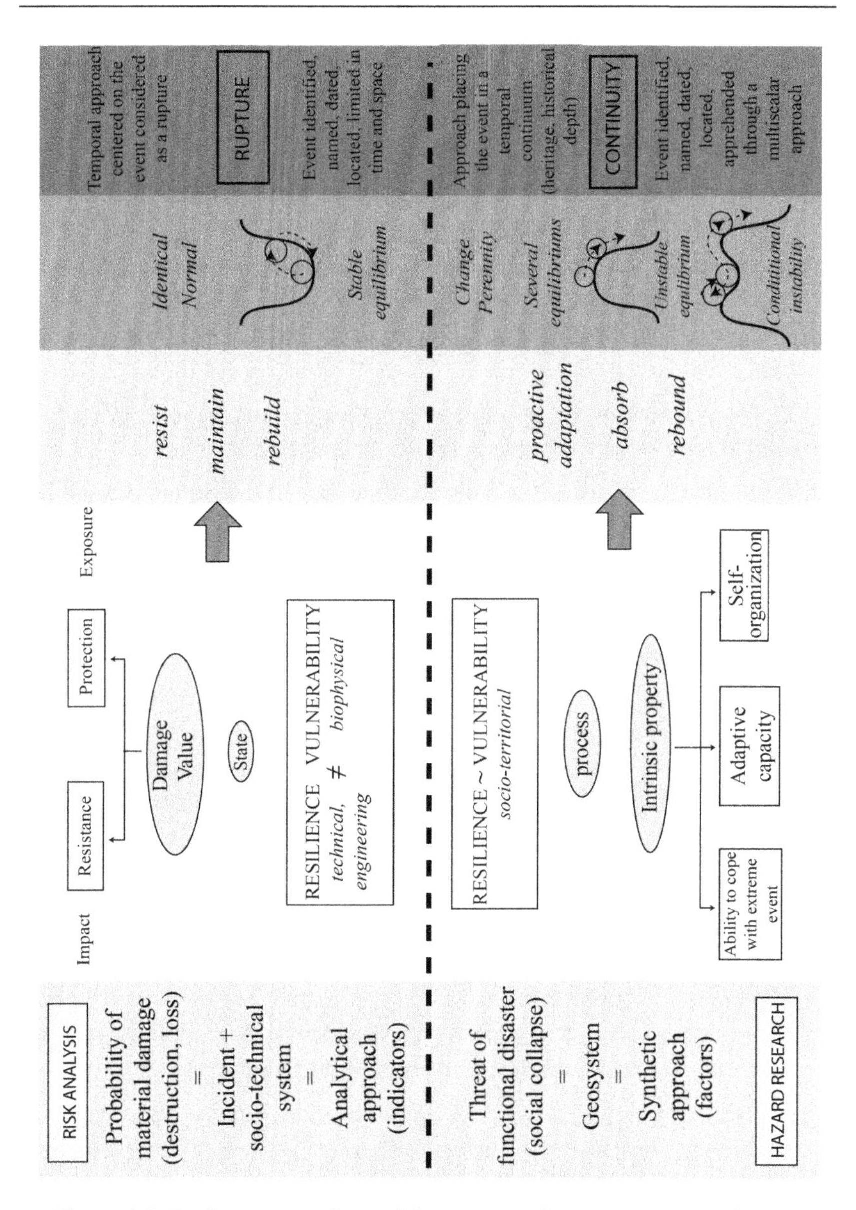

Figure 1.2. *Resilience, a single word, but two complementary concepts that are irreducible. © Magali Reghezza-Zitt & Damienne Provitolo*

1.6. Bibliography

[ADG 00] ADGER W.N., "Social and ecological resilience: are they related?", *Progress in Human Geography*, vol. 24, no. 3, pp. 347–364, 2000.

[ASC 00] ASCHAN-LEYGONIE C., "Vers une analyse de la résilience des systèmes spatiaux", *L'Espace Géographique*, no. 1, pp. 67–77, 2000.

[ANA 09] ANAUT M., "La relation de soin dans le cadre de la résilience", *Informations sociales*, no. 156, pp. 70–78, 2009.

[ARC 98] ARCHAEOMEDES, *Des oppida aux métropoles*, Paris, Anthropos, 1998.

[AUG 05] AUGENDRE M., *Risques et catastrophes volcaniques au Japon, Habiter les territoires à risques*, PPuR, Lausanne, 2011.

[BER 08] BERQUE A. (ed.), "La Ville se refait-elle?", *Géographie et cultures*, no. 65, January 2008.

[BER 10] BERTRAND F., "Changement climatique et adaptation des territoires", in ZUINDEAU B. (ed.), *Développement durable et territoire*, Villeneuve d'Ascq, Presses Universitaires du Septentrion, coll. Environnement et sociétés, pp. 339–350, 2010.

[BOW 69] BOWLBY J., "Continuité et discontinuité: vulnérabilité et résilience", *Devenir*, no. 4, pp. 7–31, 1969.

[BLA 94] BLAIKIE P., CANNON T., DAVIS I. *et al.*, *At Risk: Natural Hazards, People's Vulnerability and Disasters*, Routledge, London, 1994.

[BUR 78] BURTON I., KATES R.W., WHITE G.F., *The Environment as Hazard*, Guildford Press, New York, 1978.

[COL 10] COLLIN D., Brouillard ou arc-en-ciel de sens?, available at: http://lionel.mesnard.free.fr/le%20site/psy-resilience.html. Accessed November 15, 2010.

[COM 10] COMFORT L.K., BOIN A., DEMCHAK C., *Designing Resilience. Preparing fo Extreme Events*, University of Pittsburgh Press, 2010.

[CUT 08] CUTTER S.L. *et al.*, "A place-based model for understanding community resilience to natural disasters", *Global Environmental Change*, vol. 18, pp. 598–606, 2008.

[CYR 01] CYRULNIK B., *Les vilains petits canards*, Paris, Odile Jacob, 2001.

[CYR 06] CYRULNIK B., DUVAL P. (eds), *Psychanalyse et résilience*, Paris, Odile Jacob, 2006.

[DAU 07] DAUPHINÉ A., PROVITOLO D., "La résilience: un concept pour la gestion des risques", *Annales de Géographie*, no. 654, pp. 115–125, 2007.

[DAU 13] DAUPHINÉ A. PROVITOLO D., *Risques et catastrophes. Observer, spatialiser, comprendre, gérer*, Armand Colin, 2nd edition, 2013.

[DJA 05] DJAMENT G., La reproduction de la centralité romaine. De la Ville Éternelle à la capitale de l'Italie, PhD Thesis, University of Paris VII – Diderot, 2005.

[DOV 92] DOVERS S.R., HANDMER J.W., "Uncertainty, sustainability and change", *Global Environmental Change*, vol. 2, no. 4, pp. 262–276, 1992.

[FAB 87] FABIANI J-L., THEYS J., *La société vulnérable*, Paris, Presses de l'ENS, 1987.

[FOL 02] FOLKE C. *et al.*, Resilience and Sustainable Development: Building Adaptive Capacity in a World of Transformations, Environmental Advisory Council to the Swedish Government, Stockholm, Sweden, 2002.

[GAL 06] GALLOPIN G.C., "Linkages between vulnerability, resilience and adaptive capacity", *Global Environmental Change*, vol. 16, no. 3, pp. 293–303, 2006.

[GOD 02] GODARD O., HENRY C., LAGADEC P. *et al.*, *Traité des nouveaux risques*, Paris, Gallimard, 2002.

[GOD 03] GODSCHALK F.R., "Urban hazard mitigation: creating resilient cities", *Natural Hazards Review*, vol. 4, no. 3, pp. 136–143, 2003.

[HAR 04] HARTER H., "Chicago et l'incendie de 1871: entre mythes et réalité", in CABANTOUS A. (ed.), *Mythologies urbaines. Les villes entre histoire et imaginaire*, Presses Universitaires de Rennes, pp. 219–236, 2004.

[HER 09] HERNANDEZ J., "The long way home: une catastrophe qui se prolonge à La Nouvelle-Orléans, trois ans après le passage de l'ouragan Katrina", *Espace géographique*, vol. 38, no. 2, pp. 124–138, 2009.

[HER 10] HERNANDEZ J., ReNew Orleans? Résilience urbaine, mobilisation civique et création d'un capital de reconstruction à la Nouvelle Orléans après Katrina, PhD Thesis, University of Paris X – Nanterre, 2010.

[HEW 83] HEWITT K. (ed.), *Interpretation of Calamity from the Viewpoint of Human Ecology*, Allen and Unwin, London, 1983.

[HOL 73] HOLLING C.S., "Resilience and stability of ecological systems", *Annual Review of Ecology and Systematics*, no. 4, pp. 1–23, 1973.

[HOL 01] HOLLING C.S., "Understanding the complexity of economic, ecological and social systems", *Ecosystems*, no. 4, pp. 390–405, 2001.

[KLE 03] KLEIN R.J.T., NICHOLLS R.J., THOMALLA F., "Resilience to natural hazards: how useful is the concept?", *Environmental Hazards*, vol. 5, nos. 1–2, pp. 35–45, 2003.

[KNI 06] KNIGHT F.H., *Risk, Uncertainty and Profit*, Dover Publication, New York, 2006.

[LAB 09] LABEUR C., "Les formes d'organisation spontanée et l'entraide au cours des catastrophes: le cas des inondations dans le delta du Rhône", in BUCHET L., RIGEADE C., SÉGUY I. *et al.* (eds), *Vers une anthropologie des catastrophes*, Antibes, Paris, APDCA/INED, pp. 221–236, 2009.

[LAG 93] LAGADEC P., *Apprendre à gérer les crises: société vulnérable, acteurs responsables*, Paris, Editions d'Organisation, 1993.

[LAL 08] LALLAU B., "La résilience, moyen et fin d'un développement durable?", *Ethics and Economics*, vol. 8, 2008.

[LHO 12] LHOMME S., Les réseaux techniques comme vecteur de propagation des risques en milieu urbain. Une contribution théorique et pratique à l'analyse de la résilience urbaine, PhD Thesis, University of Paris Diderot and Paris-Est (EIVP), 2012.

[MUS 02] MUSSET A., *Villes nomades du nouveau monde*, Paris, Éditions de l'École des hautes études en sciences sociales, p. 397, 2002.

[PIE 08a] PIERDET C., Les temporalités de la relation ville–fleuve à Phnom Penh (Cambodge) – La fixation d'une capitale fluviale par la construction d'un système hydraulique (1865–2005), PhD Thesis of Geography, University of Paris I – Panthéon-Sorbonne, November 2008.

[PIE 08b] PIERDET C., "Prévoir la trajectoire d'une cité fluviale endiguée. Mise en crise et résilience du système hydraulique de Phnom Penh (Cambodge) depuis les années 1960", in GRATALOUP C., MARTIN P. (eds), *Géopoint 2006*, Groupe Dupont et UMR Espace, Université d'Avignon, pp. 191–195, May 2008.

[PIM 84] PIMM S.L., "The complexity and stability of ecosystems", *Nature*, no. 307, pp. 321–326, 1984.

[PRO 09] PROVITOLO D., *Vulnérabilité et résilience, géométrie variable de deux concepts*, available at http://hal.archives-ouvertes.fr/hal-00497757/fr/, 2009

[QUA 05] QUARANTELLI E.L., PERRY R.W., *What is a Disaster: New Answers to Old Questions*, Xlibris Corporation, p. 444, 2005.

[QUE 13] QUENAULT B., "Retour critique sur la mobilisation du concept de résilience en lien avec l'adaptation des systèmes urbains au changement climatique", *EchoGéo*, vol. 24, 2013.

[REG 06] REGHEZZA M., Réflexions sur la vulnérabilité métropolitaine. La métropole parisienne face au risque de crue centennale, PhD Thesis, University of Paris X – Nanterre, 2006.

[REG 12] REGHEZZA M., RUFAT S., DJAMENT G. *et al.*, "What resilience is not: uses and abuses", *Cybergeo*, no. 621, 2012.

[ROB 13] ROBERT S., NOIZET H., "L'étude de la résilience des formes entre archéologie, histoire et géographie", available at http://www.geographie.ens.fr/Journee-d-etudes-20-avril-2013.html, 2013.

[SEN 85] SEN A.K., *Commodities and Capabilities*, Amsterdam & New York, North-Holland, 1985.

[SER 13] SERRE D., BARROCA B., LAGANIER R., *Resilience and Urban Risk Management*, CRC Press, Taylor & Francis Group, London, UK, 2013.

[TIM 81] TIMMERMAN P., *Vulnerability: Resilience, and the Collapse of Society: A Review of Models and Possible Climatic Applications*, Environmental Monograph, no. 1, 1981.

[TOU 12] TOUBIN M., LHOMME S., DIAB Y. *et al.*, "La Résilience urbaine: un nouveau concept opérationnel vecteur de durabilité urbaine?", *Développement durable et territoire*, 2012.

[VAL 05] VALE L.V., CAMPANELLA T.J. (eds), *The Resilient City. How Modern Cities Recover from Disaster*, Oxford University Press, New York, 2005.

[VAL 09] VALLAT C. (ed.), *Pérennité urbaine, ou la ville par-delà ses métamorphoses*, Paris, LHarmattan, 2009.

[WER 89] WERNER E.E., "Children of the garden Island", *Scientific American,* no. 4, pp. 76–81, 1989.

[WIL 98] WILDAVSKY A., *Searching for Safety*, Transaction Books, New Brunswick, N. J., 1998.

[WIS 76] WISNER B., O'KEEFE P., WESTGATE K., "Taking the naturalness out of natural disaster", *Nature*, vol. 260, no. 5552, pp. 566–567, 1976.

2

Resilience and Vulnerability: From Opposition towards a Continuum

The previous chapter has provided a reminder of the polysemy of the term "resilience" and showed how resilience has become prominent in the field of disaster management, where vulnerability is perceived as a less fundamental concept as a result. However, most definitions of resilience share the view of a relationship with definitions of vulnerability [BAR 13], in such a way that it seems both concepts can only be understood in relation to each other [VAN 01]. Most often, vulnerability is reported as something rather negative, and resilience as something positive.

The highlight of the "positive" character of resilience can be explained largely by a search for applicability: while vulnerability refers to weakness and incapacity, resilience is deemed as the desirable property of a system, toward which management is likely to tend [KLE 03]. This leads to a systematic opposition between the two concepts. However, this vision of the vulnerability/resilience pair is problematic from a theoretical point of view.

This chapter has a twofold objective. On the one hand, we will shed some light on the vulnerability/resilience pair through the analysis of works carried out in the fields pertaining to risks, disasters and climate change. On the other hand, we put forward a new

---

Chapter written by Damienne PROVITOLO and Magali REGHEZZA-ZITT.

methodological and conceptual framework to rethink the relationship between the two concepts from the notion of "resiliencery vulnerability" [PRO 12].

2.1. One or several vulnerabilities?

The polysemy of the term "resilience" refers to the concept of vulnerability. There are three complementary approaches for the analysis of vulnerability, in which each approach is linked to different paradigms. One of them, the oldest one, analyzes vulnerability from the measurement of real or potential damage to the elements threatened by a disaster. The second one looks at vulnerability from the perspective of society's response capacity in the face of a disturbance. Finally, the third and most recent approach focuses on territorial issues. In the first case, the appropriate term is biophysical vulnerability, in the second one, it is social vulnerability, and in the third case, it is territorial vulnerability [BON 02].

2.1.1. *Biophysical vulnerability*

Biophysical vulnerability encompasses notions developed in physics and engineering sciences starting in the 1950s: fragility, susceptibility, sensitivity, damage, etc. It measures the physical impact of a hazard on buildings, networks, infrastructures and populations, called "elements at risk" (hence the prefix bio-, which aims at integrating non-inert components). According to the authors, it can refer not only to the damage actually suffered, but the expected damage as well. This damage is assessed either absolutely or relatively.

Biophysical vulnerability depends on three factors: the sensitivity of the element at risk, its physical resistance and its exposure to the hazard [ADG 00]:

– *sensitivity* is a function of the nature, severity and frequency of the damaging event. Thus, it is peculiar to the hazard: the Kobe earthquake of 1995, as an example, showed that old wooden houses, whose roofs were burdened to resist cyclones, as a result, were made more sensitive to earthquakes [MEN 01];

– *resistance* was, for a long time, reduced to the physical resistance of buidings and protection infrastructures, and was later extended to the physical and/or mental resistance of one individual or any group. This accepted sense has led managers to develop a series of technical solutions aiming at increasing this resistance: for instance, building standards, such as earthquake-resistance standards and sealing standards for some materials, have gradually become mandatory in developed countries;

– *exposure* reflects the part of the proximity to the potential hazard in damage mechanisms. Some authors state vulnerability is the measurement of the degree of exposure [ALE 93]. This definition helps risk assessment mapping, by overlaying maps of the spatial extension of the potential hazard and the distribution of elements at risk [PIG 05]. It calls for solutions that rely on constructing barriers and controlling the landuse in risk areas, especially by implementing the regulatory provisions of zoning.

However, this approach underestimates the social dimension of vulnerability, in particular the diversity of the response capacities of societies facing hazards and disasters.

2.1.2. *Social vulnerability*

Social vulnerability supplements the previous approach. It covers the capacity for damage in terms of responses, meaning the active adaptation of the elements at risk. It shifts the focus from the physical process – an external risk factor – toward the ability to cope with the threat and with the impact of the damaging process.

While biophysical vulnerability is first considered as a state, social vulnerability is defined as a quality and a property of the element at risk that explains this state. At least at the start, it has a certain heuristic value: it is less about measuring the potential for damage than revealing factors that generate this potential, by focusing on endogenous risk factors that are not necessarily specific to the hazard.

These factors are of various types: political, institutional, economic, sociocultural, psychological and cognitive, etc. [FAB 87]. They involve everything that enables an individual, a group, an institution or a society to increase or reduce its ability to anticipate an event, cope with it, manage it and, for some authors, overcome it. However, the framework for the analysis of social vulnerability has little to do with the geographical context and territorial issues. To deal with this limitation, a new approach has been suggested. This is territorial vulnerability.

2.1.3. *Territorial vulnerability*

Territorial vulnerability is a recent concept, developed by French geographers. There is not such a concept of territory in the English-speaking world. The term "territory" became prominent in the 2000s in French public publications[1]. Territorial is here synonymous with "local", as a "sign of the adaptation of these policies to local contexts and stakes, which is meant to ensure democratic readability and socio-economic efficiency" [GHI 06].

French geographers have shown that the "spatialization" of risk [NOV 02], that is to say spatially delineating risk areas, is the basis of their management policy [BEU 08]. Spatial delineation, which gives rise to operation management perimeters and zoning, does rely on scientific knowledge, but especially on political grounds, sources of conflict, that can be analyzed due to the concept of territory when the latter is marked by a strong geopolitical dimension [PIG 05].

Geographers also address impacts of the definition of a given risk in a given space. They highlight, for example, that the risk can "create new spatial baselines, new types of territorial anchoring that standard administrative frameworks no longer match" [BEU 05]. Labeling a given space as an area "at risk" is not unbiased. The identity and image of the area, as a result, become altered necessarily. Risk helps

1 For example, in 2003, the review *Pouvoirs Locaux* published a special issue dedicated to the territorialization of risk as an instrument for better risk management.

generate to "territory", this time understood as an assigned geographical area.

Some researchers go even further and put forward the hypothesis that space is a "structure likely to cause disruptions in and disturbances of itself" [NOV 94]. Particularly, some spatial configurations may generate, exacerbate or mitigate risks [REG 09]. Whatever the nature of the initial damaging process, risk dynamics therefore become specific to the affected territory. Hence, the risk may be given as urban or metropolitan, meaning that a same hazard (for example a natural or technological hazard event) would have specific impacts depending on the area where it occurs, for each type of space has its own unique vulnerabilities [REG 06]. Therefore, studying territorial vulnerability implies identifying not only areas likely to suffer major damage, but locations from which disturbances may spread away as well [DER 09], and spatial layouts able to help prevent that spread [CUT 00].

Territorial vulnerability is therefore an attractive yet ambiguous concept. The territorial approach mainly provides a synthetic view which proceeds beyond the dichotomous one. The risk turns out to be an element outside of any territory, so that the issue "is no longer about how manage the risk, once identified on a territory, but how manage one territory including one or more potential risks". [BEU 08]".

Starting with a territory assumes the rejection of the disaggregated approach to risks which would separate the hazard event, vulnerability, and its various components, from one another [PIG 05]. Regarding the territory as a system, research works on territorial vulnerability are getting closer to those on systemic vulnerability, with these works focusing on domino and interdependence effects in particular [PRO 02].

2.2. The vulnerability/resilience pair

To report the complexity of vulnerability, some researchers have developed a systemic approach [PIG 05], suggesting a cross-cutting and holistic vision.

The Intergovernmental Panel on Climate Change (IPCC) complies with the trend towards defining vulnerability as:

> *The degree to which a system is susceptible to, and unable to cope with, adverse effects of climate change, including climate variability and extremes. Vulnerability is a function of the character, the magnitude, and rate of climate change and variation to which a system is exposed, its sensitivity, and its adaptive capacity* [CAR 94].

The systemic approach to vulnerability has shown it has a twofold dimension: it is both endogenous to the system (as a state of fragility) and a function of its ability to encounter a disturbance, absorb it, adapt to it and return back to normal functioning [DAU 13]. The notion of systemic vulnerability, because it refers to adaptation capabilities, opens the path towards resilience.

2.2.1. *Resilience as the opposite of vulnerability*

The first way to consider the vulnerability/resilience pair is to oppose them as the two sides of a coin [FOL 02]. By that way, the group *Resilience Alliance*, whose goal is to stimulate multidisciplinary research on resilience and inform international policy makers, defines it as the "flip-side" of vulnerability.

Opposing vulnerability and resilience is a logical consequence of the definition of resilience as a capacity to adapt to impact, and those of vulnerability as a degree of sensitivity to damage. If any element at risk or any system suffers some damage, then it is vulnerable. The more damaged it becomes the higher the probability it disrupts and falls apart, and the more difficult it is to recover. In this way, the more vulnerable the system is (in the sense of being sensitive to damage), the less resilient it is.

The opposition between vulnerability and resilience is equally obvious if we no longer consider the impact of the disturbance, but rather the incapacity to cope with it. The system may fall apart as it cannot anticipate nor adapt to various threats. It is even less capable of withstanding and absorbing one of these threats. All the definitions of resilience as seen in the previous chapter are implicitely reminded.

The systematic opposition between vulnerability and resilience leads to deduce that reducing vulnerability means increasing resilience mechanically. R. Klein *et al.* [KLE 03] define five ways of improving the ability to adapt to climate change (and resilience consequently), which are, in fact, all measures for reducing social and biophysical vulnerability: increase the resistance capacity of buildings; increase the flexibility of anthropogenic systems by modifying the management strategy uses and locations; develop the adaptability of natural systems; reverse dynamics leading to an increase in vulnerability (in particular human settlements in risk areas); improve public awareness about potential risks and provide populations with adequate information.

However, experience has shown that being both vulnerable and resilient to impacts is a possible feature. With regard to cities, their potential for damage is tremendous, and some authors claim it has never been so great [MIT 99]. The level of vulnerability of cities is high, because of their exposure to various threats, and their own fragility as well. However, cities are at the same time highly resistant to disasters since, as highlighted by L.J. Vale and T.J. Campanella:

> *Although cities have been destroyed throughout history – sacked, shaken, burned, bombed, flooded, starved, irradiated and poisoned – they have, in almost every case, risen again like a phoenix. [...] Cities are among humankind's most durable artefacts* [VAL 05].

Cities offer a large potential for recovery, due to the concentration of labor, wealth and innovation capacities that shape the very basis of the term "urban". Moreover, they can often rely on their hinterlands for support. Their strategic value means they concentrate strengths for relief work and recovery. They are therefore both vulnerable and resilient, as some of their own features may emerge as a vulnerability and/or resilience factor, depending on the time. For instance concentrations of human, technical and financial capitals not only promote damage (therefore biophysical vulnerability) but also help overcome the crisis (and therefore resilience).

Vulnerability even causes resilience, since strength derives from vulnerability in the strictest sense, resilience arises only in case of some impact or disturbance, which implies vulnerability does exist. To take that reasoning to its logical conclusion, the following paradox may be underlined: the more a system is subjected to the consequences of its vulnerability, the more it undergoes crises, and the more it is able to prove its resilience capacities (property), generate positive feedback, and therefore actually become resilient (as a state/result).

2.2.2. *Resilience as a component of vulnerability*

Consequently, the relationships between vulnerability and resilience are far from being as clearly stated as the "flip-side" images suggest. Scientific literature is often ambiguous to that concern. Folke *et al.* define resilience as being the counterpart of vulnerability, but on the same page, they make it one of the three defining elements of vulnerability, alongside sensitivity and exposure [FOL 02]. S. Cutter *et al.* define social vulnerability as the possibility for a group to suffer losses or as its resilience to a hazard (risk perception, prevention, experience, etc.):

> *Social vulnerability is partially a product of social inequalities—those social factors and forces that create the susceptibility of various groups to harm, and in turn affect their ability to respond, and bounce back (resilience) after the disaster* [CUT 03].

As a result, the capacity for resilience becomes a component of vulnerability, as the ability to stand up to threats, although with certain degrees of variation included. For B. Quenault:

> *Some authors have represented these various concepts as being set within the greater concept of vulnerability [TUR 03, GAL 06]; others have emphasized that resilience, which according to them is completely included in the capacity for adaptation, is one of the components of vulnerability* [ADG 06, BIR 06, CUT 08].

2.3. Beyond opposition: the notion of "resiliencery vulnerability"

The complexity of the relationships between resilience and vulnerability is undoubted: while both concepts are not entirely independent of each other, they are distinct. From our point of view, their relationship cannot be reduced to either strict opposition or even inclusion: there is a *continuum* between the two of them, meaning that vulnerability can be intertwined with and modified by resilience considered from a wider perspective [PRO 12].

To reflect this view, we propose the notion of "resiliencery vulnerability" (a neologism that we have created on purpose) and a conceptual framework to analyze it. Our objective is threefold: question the opposition between, on the one hand, vulnerable systems, and on the other hand, resilient systems; show that resilience and vulnerability can be contingent; stress the fact that resilience is not inherently positive–it may have negative effects.

The model illustrated by Figure 2.1 allows for a global analysis of the "resiliencery vulnerability" of socioeconomic and territorial systems. We can differentiate the factual part, which formalizes the context of risk, from the "resiliencery vulnerability" part, which describes potentialities, capacities and reactions of a system to protect itself against risk or disaster. The analysis covers a complete temporal environment: before, during and after the disturbance or damaging event.

2.3.1. *Explanation of the factual part of the model*

The factual part identifies the information needed for the analysis of a system submitted to a context of risk [PRO 14]. It can be split into three main sets:

– elements of the system, which can be of a various types;

– events that threaten this system: they can be minor or major, natural, industrial, technological or social, point source or non-point source;

– damage caused by disaster events of various types and extents.

The generic term "element" includes the "living element", "physical element" and "miscellaneous infrastructure" that make up the studied system. This classification enables to encompass all components of the system that can be involved in situation of a risk, an accident, a disaster or a catastrophe. The "living element" includes all humans and natural populations, such as fauna and flora, in other words the biological components of the system. The "physical element" refers to the abiotic components of geosystems (landscape, climate, water bodies, grounds, etc.). Finally, the "miscellaneous infrastructure" includes buildings, facilities, networks, etc., or, in other words, man-made structures and items that have shaped the territory.

The "event" itself may have occurred (disaster), or be likely to happen or be potential (risk). It therefore refers to the threat of destruction or disturbance as well as a true occurrence. In general, the event is exceptional and striking compared to everyday life. However, some events, due to their recurrent nature, are characterized by a potential that is catastrophic without being exceptional: droughts in Africa, flooding due to monsoons in Asia, earthquakes along the Ring of Fire, flash floods in the Mediterranean urban areas or cyclones in the inter-tropical zone.

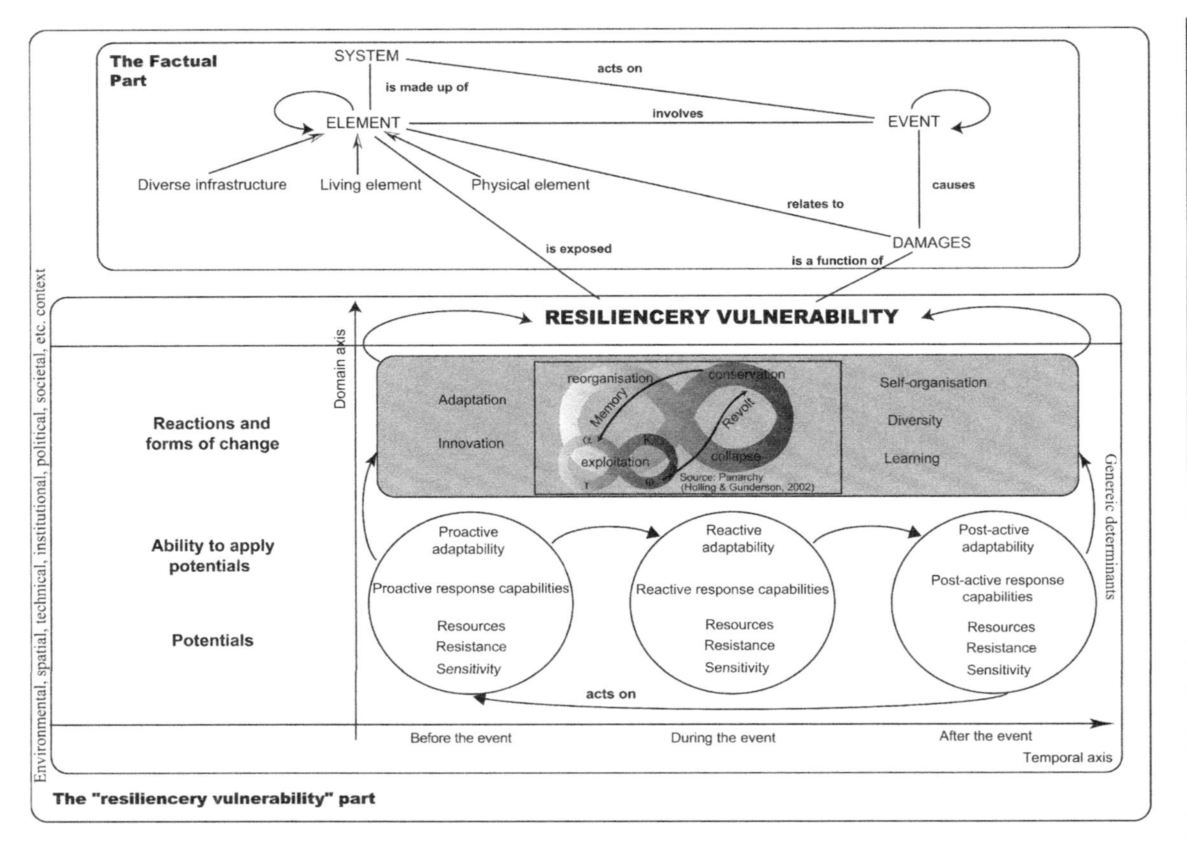

Figure 2.1. *A conceptual model for the analysis of "resiliencery vulnerability" © Damienne Provitolo*

We should note that an event can affect one or several elements of the system: for instance, pollution can destroy an ecosystem without disrupting infrastructures or socioeconomic components; a pandemic can affect the living elements but not infrastructures.

According to the model, the "brings into play" relationship joins events and elements together. For example, the economic downturn of a city or country can contribute to worsen the effects of a natural hazard process, and technical networks can hasten the spread of a disturbance, etc. Moreover, this "brings into play" relationship indicates that the element is exposed to the event, but this exposure does not presuppose its vulnerability: A vaccinated individual might be in contact with the pandemic without running the risk of any damage (or at least with a significantly mitigated risk); a building that complies with earthquake-resistance standards would suffer little damage in the event of an earthquake. Finally, an event can also cause other events: a natural hazard can generate pollution, outages, epidemic risks, etc. Then, events are interrelated by relationships of causality and interdependence (domino effect).

2.3.2. *Explanation of the "resiliencery vulnerability" part of the model*

The "resiliencery vulnerability" part of the model is structured around two axes: the temporal axis and the domain axis.

2.3.2.1. *The temporal axis*

The temporal axis enables to understand "resiliencery vulnerability" as a dynamic process. The model articulates the pre-event time (implementation of preventive strategies allowing, *in fine*, to avoid or mitigate the outbreak of a crisis in the system), the present time (occurence of the event that generates a crisis situation) and the future (the post-event). These three temporal environments affect each other but cannot be considered at the same time since they are time-ordered events. The dynamics at work are complex due to the emergence of retroaction loops [PRO 09] with temporally shifted effects.

Preventive measures aim at reducing biophysical and social vulnerability, and, to a lesser extent, the territorial vulnerability of a system before an event occurs. These measures affect the progress of the crisis operations against the hazard mitigate its potential for disturbance; strengthening building codes can reduce material damage, and, as a result, the functional disturbance that might have arisen from this potential damage; controlling exposure reduces the number of elements that are affected, thus lessening the number of people to be rescued or evacuated; dissemination of preventive information leads to appropriate behaviors, preventing panics. Following the event, a post-disaster analysis, better known as an experience feedback, helps to identify and analyze the failures and errors in the crisis management, but during the prevention upstream phase as well. New strategies are then developed, including lessons learned from the past.

2.3.2.2. *The domain axis*

The domain axis illustrates the potentialities, capacities and reactions of a system before, during and after its occurrence. This formalization sets up a differentiation between what is available (potentialities), what can be done (capacities) and what is actually done (reactions and forms of change). In fact, reactions are limited by capacities (what we can do) and potentialities (what is available) and capacities become actual capacities when they match the system reactions. Each phase therefore interacts with the others.

The potentialities of a system refer to the resources and capital available (economic, social, human, knowledge-based, technical, financial, etc.). These resources may be inherent to the system, or they can provide from its more or less immediate environment, especially after the event. In 2010 for instance, the Haitians were able to rely on international aid and on the diaspora resources as well.

Capacities enables the system to mobilize its potentialities, use its resources and the available capital and to get profit from them. Nevertheless it is not because a resource is available that it is used properly or even at all. In 2010, the cyclone Xynthia showed that, despite the presence of efficient predictive devices, the warning did not work properly. Capacities refer to the adaptive capacity of the system, understood here as the capacity to face and respond to the threat, emergency or disorganized state that follows the crisis [BUR 02]. The adaptive capacity is qualified as proactive, reactive or post-active, depending on whether it is expressed before, during or after the event. Actually, it would be more appropriate to talk about adaptability (a term that we favor in the model); capacities are more representative of a potential for adaptation, rather than effective adaptation [BRO 03].

Finally, *reactions* correspond to solutions that are implemented within a system undergoing a crisis, while *forms of change* refers more to the different trajectories through which the system can pass. Reactions and forms of change are linked. They depend on the existing resources and on the effective capacity to mobilize them at the right moment and in an adequate manner.

2.3.2.3. *The panarchy model and the introduction of systemic resilience*

The reaction phase involves the panarchy model of Gunderson and Holling [GUN 02] (Figure 2.2). This model describes the dynamics of a system exposed to a disturbance as it switches from a state of stability to another. Its originality lies in its proposal for a multiscalar analysis of resilience, from a systemic perspective. It relies on the notion of an adaptive cycle, with four phases that ensure the transition between different states of stability: exploitation, conservation, release and reorganization.

Exploitation is a phase of resilience during which the system can absorb high disturbances. Conservation is characterized by a major state of stability, with low levels of resilience to the disturbances. Release is a phase in which the structure built during the previous phases becomes disorganized. The reorganization phase, stressed by high levels of instability, can either lead to the initial state of stability, or to a new adaptive cycle.

The connections between the different levels of adaptive cycles are carried out along two paths:

– the first path, called "revolt", represents the passage from the release phase (φ) to the conservation phase (K): it expresses the situation when fast events, at a lower scale, overrun slow processes at a higher scale. For example, the attacks of September 11 in 2001 accelerated the process of recomposure of the economic activities of Manhattan (slower process) and altered New Yorks's trajectory (higher scale) [SAS 02];

– the second path, called "remember", represents the direct passage from the conservation phase (K) to the reorganization phase (α) by mobilizing heritages (biological, institutional and economic). As stated by Gunderson and Holling [GUN 02], following a fire event, the renewal of a forest relies on physical structures or surviving species, which make up the heritage of the growth phase of the forest. It also mobilizes resources on larger scales (role of insects and birds in the transportation of seeds, for example)

The panarchy model therefore uses a multiscalar approach. The disappearance of a subsystem can thus put forward and feed the capacity for resilience of a meta-system. For example, the subway network (meta-system) may maintain its service despite a station (subsystem) not being rebuilt following a terrorist attack or a flood.

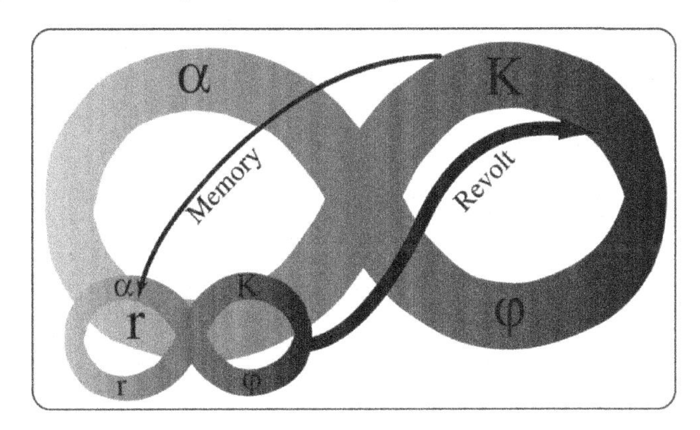

Figure 2.2. *The panarchy model [GUN 02]*

In the context of the model of "resiliencery vulnerability", the panarchy model is coupled with factors that determine the dynamics of a system and therefore its trajectories. Five factors are often mentioned: diversity, self-organization, learning, adaptation and innovation [ADG 05].

2.3.3. *The vulnerability/resilience pair in the "resiliencery vulnerability" model*

The "resiliencery vulnerability" model highlights the tight links between resilience and vulnerability and shows that relationships between the two concepts are ambivalent.

For example, adaptability is seen as a property that increases resilience. However, strong adaptability can, unintentionally, lead to a loss of resilience, and this is true for at least three cases [WAL 06]:

– the adaptability of some groups can come at the expense of other groups, and micro-scale adaptability can come at the expense of macro-scale adaptability (and vice versa). A company can, for example, face a crisis by stopping work with its subcontractors: its adaptability threatens the survival of lower level companies and disorganizes the economic system that exists at a regional level;

– adaptability to a known impact can generate increased vulnerability to unknown or extreme impacts. For example, collective protective structures (levees, dams, etc.) can minimize damages and enable a rapid recovery of the system. However, the cyclone Xynthia in February 26–28, 2010 showed that these defensive works are vulnerable to extreme events and give populations a false sense of security;

– finally, adaptability can lead to a loss of the diversity of responses, when response strategies rely more on the anticipation of known and predictable events than on the handling of unknown and unpredictable events.

Furthermore, depending on the sub-field (property/capacity/reaction) and depending on the spatial and temporal scales considered, situations can be interpreted either in terms of vulnerability, or in terms of resilience. For instance, the vulnerability of the material infrastructure of a transportation network (property before the event) does not have to lead to a stoppage of the provided service (reaction during the event). The structure of the network and behaviors of the network managers (capacities), which rely on crisis management and service continuity plans (resources made available before the event), can ensure the service continuity and thus the resilience of infrastructure functions. We can therefore consider that the system is either vulnerable or resilient: it all depends on which notion we want to stress.

The model also highlights that vulnerability and resilience are not interconnected like connecting vessels. The mobilization of technical resources in order to increase the capacity for resistance of infrastructures *before* the event thus decreases the biophysical vulnerability of the system *during* the event and increases the system's potential for resilience *after* the event. A decrease in vulnerability leads here to an increase in resilience. However, if the vulnerability of an element of the system, revealed *during* the event, decreases the potential for resilience of the system *after* the event, it can as well be necessary to reveal a weakness of the system that is *prior to* the event and to allow an increase in resilience in the *longer term*. For example, in France, the storm of 1999 revealed the vulnerability of forest ecosystems [WIK 15]. However, the storm also raised awareness concerning the failings in the management of French forests, and led to a resetting of some devices and policies which should ensure greater resilience of forests in the long term. Thus, vulnerability decreases resilience in the short term, but increases it in the long term. Furthermore, this example shows that resilience cannot imply a return to an earlier state, which would maintain the vulnerabilities of the system and of its components.

Even more paradoxically, the model of "resiliencery vulnerability" shows that the maintenance of some vulnerabilities could be necessary for the resilience of the system. Up to a certain threshold, remaining suffering impacts and damages (signs of a certain degree of vulnerability) can actually be considered as a resource (*potential*), as it facilitates the system to stay on the alert and to maintain its *capacities* and its *reactions*: the recurrence low magnitude earthquakes in Japan, for example, prepares it for the occurence of the "Big One".

Moreover, the vulnerability of some elements of the system can become a factor of the resilience of the system as a whole: flood spillways maintain the vulnerability of some spaces that shall be flooded so as to ensure the resilience of the system at the macro level. Just as resilience can be regarded as negative, vulnerability can be considered as a positive attribute when it allows forms of vulnerability to continue while leaving systems unchanged.

2.4. Conclusion

The model of "resiliencery vulnerability" formalizes the intertwining of vulnerability and resilience, and highlights the interactions between the two notions. Through this neologism, resilience is directly connected to vulnerability, but not specifically as a form of opposition. The idea of a *continuum* is the one put forward.

The model also allows us to understand how resilience can have a positive or negative effect, depending on the scale of the analysis of the system, the nature of the risk and its severity. "Resiliencery vulnerability" does not negate the fragility of the system studied, but it does highlight all of the potentialities, capacities and reactions that will allow it, despite this fragility, to deal with minor events as if they were extreme. The model also reiterates the fact that living systems do not remain passive when faced with events. This is why, the notion of "resiliencery vulnerability" is dynamic.

Finally, this "resiliencery vulnerability" varies over time, and from one country to another, depending on geographical, spatial, institutional, political, social and economic contexts: it integrates the various approaches for vulnerability as mentioned in the first part, and especially territorial approaches.

2.5. Bibliography

[ADG 00] ADGER W., "Social and ecological resilience: are they related?", *Progress in Human Geography*, vol. 24, pp. 347–364, 2000.

[ADG 05] ADGER W.N., HUGHES T.P., FOLKE C. *et al.*, "Social-ecological resilience to coastal disasters", *Science*, vol. 309, no. 5737, pp. 1036–1039, 2005.

[ADG 06] ADGER W.N., "Vulnerability", *Global Environmental Change*, vol. 16, no. 3, pp. 268–281, 2006.

[ALE 93] ALEXANDER D.E., *Natural Disasters*, Chapman & Hall, New York, 1993.

[BAR 13] BARROCA B., DINARDO M., MBOUMOUA I., "De la vulnérabilité à la résilience: mutation ou bouleversement?", *EchoGéo*, no. 24, available at http://echogeo.revues.org/13439, 2013.

[BEU 05] BEUCHER S., REGHEZZA M., *Les risques*, Paris, Bréal, 2005.

[BEU 08] BEUCHER S., MESCHINET DE RICHEMOND N., REGHEZZA M., "Les territoires du risque. Exemple du risque inondation", *Historiens et géographes*, no. 403, pp. 103–111, 2008.

[BIR 06] BIRKMANN J., "Measuring vulnerability to promote disaster-resilient societies: conceptual frameworks and definitions", in BIRKMANN J. (ed.), *Measuring Vulnerability to Natural Hazards – Towards Disaster Resilient Societies,* United Nations University Press, Hong-Kong, New York, pp. 7–54, 2006.

[BON 04] BONNET E., Risques industriels: évaluation des vulnérabilités territoriales, PhD Thesis, University of Le Havre, p. 341, 2002.

[BRO 03] BROOKS N., "Vulnerability, risk and adaptation: a conceptual framework", Tyndall Center for Climate Change Research, Working paper 38, p. 20, 2003.

[BUR 02] BURTON I., SALEEMUL H., LIM B. *et al.* "From impacts assessment to adaptation priorities: the shaping of adaptation policy", *Climate Policy*, vol. 2, nos. 2–3, pp. 145–159, 2002.

[CAR 94] CARTER T.R., PARRY M.L., NISHIOKA S. *et al.* (eds), IPCC "technical guidelines for assessing climate change impacts and adaptations", University College London and Centre for Global Environmental Research. London, UK and Tsukuba, Japan, 1994.

[CUT 00] CUTTER S.L., MITSHELL J.T., SCOTT M.S., "Revealing the vulnerability of people and places: a case study of Georgetown country, South California", *Annals of the Association of American Geographers*, vol. 90, no. 4, pp. 713–737, 2000.

[CUT 03] CUTTER S.L., BORUFF B.J., SHIRLEY W.L, "Social vulnerability to environmental hazards", *Social Science Quarterly*, vol. 84, no. 1, pp. 242–261, p. 243, 2003.

[CUT 08] CUTTER S.L., BARNES L., BERRY M. *et al.*, "A place-based model for understanding community resilience to natural disasters", *Global Environmental Change,* vol. 18, pp. 598–606, 2008.

[DAU 13] DAUPHINÉ A, PROVITOLO D., *Risques et catastrophes. Observer, spatialiser, comprendre, gérer*, Paris, Armand Colin, 2013.

[DER 09] D'ERCOLE R., METZGER P., "La vulnérabilité territoriale: une nouvelle approche des risques en milieu urbain", *Cybergeo: European Journal of Geography*, available at http://www.cybergeo.eu/index22022.html, 2009.

[FAB 87] FABIANI J.-L., THEYS J., *La société vulnérable*, Paris, Presses de l'ENS, 1987.

[FOL 02] FOLKE C. *et al.*, Resilience and Sustainable Development: Building Adaptive Capacity in a World of Transformations, Environmental Advisory Council to the Swedish Government, p. 73, Stockolm, Sweden 2002.

[GAL 06] GALLOPIN G.C., "Linkages between vulnerability, resilience, and adaptive capacity", *Global Environmental Change*, vol. 16, pp. 293–303, 2006.

[GHI 06] GHIOTTI S., "Les Territoires de l'eau et la décentralisation. La gouvernance de bassin versant ou les limites d'une évidence", *Territoires et développement durable*, available at http://developpementdurable.revues.org/document1742.html, 2006.

[GUN 02] GUNDERSON L., HOLLING C.S. (eds), *Panarchy: Understanding Transformations in Human and Natural Systems*, Island Press, Washington D.C., 2002.

[KLE 03] KLEIN *et al.*, "Resilience to natural hazards: how useful is the concept?", *Environmental Hazards*, vol. 5, nos. 1–2, pp. 35–45, 2003.

[LHO 12] LHOMME S., Les réseaux techniques comme vecteur de propagation des risques en milieu urbain. Une contribution théorique et pratique à l'analyse de la résilience urbaine, PhD Thesis, Universités Paris Diderot et Paris-Est (EIVP), 2012.

[MEN 01] MENONI S., "Chains of damages and failures in a metropolitan environment: some observations on the Kobe earthquake in 1995", *Journal of Hazardous Materials*, vol. 86, pp. 101–119, 2001.

[MIT 99] MITCHELL J.K. (ed.), *Crucibles of Hazard: Disasters and Megacities in Transition*, United Nations University Press, Tokyo, New York, Paris, 1999.

[NOV 94] NOVEMBER V., "Risques naturels et croissance urbaine: réflexion théorique sur la nature et le rôle du risque dans l'espace urbain", *Revue de Géographie Alpine*, no. 4, pp. 113–123, 1994.

[NOV 02] NOVEMBER V., *Les territoires du Risque. Le risque comme objet de réflexion géographique*, Peter Lang, Berne, 2002.

[PIG 05] PIGEON P., *Géographie critique des risques*, Paris, Economica, 2005.

[PRO 02] PROVITOLO D., Risque urbain, catastrophes et villes méditerranéennes, PhD Thesis, University of Nice Sophia Antipolis, 2002.

[PRO 09] PROVITOLO, "A new classification of catastrophes based on complexity criteria", in AZIZ-ALAOUI M.A., BERTELLE C. (eds), *From System Complexity to Emergent Properties*, Springer-Verlag, pp. 179–194, 2009.

[PRO 12] PROVITOLO D., "The contribution of science and technology to meeting the challenge of risk and disaster reduction in developing countries: from concrete examples to the proposal of a conceptual model of 'resiliencery vulnerability'", BOLAY J.-C. *et al.* (eds), *Technologies and Innovations for Development*, Springer-Verlag, 2012.

[PRO 14] PROVITOLO D., DUBOS-PAILLARD E., MÜLLER J.-P., "Une ontologie conceptuelle du domaine des risques et catastrophes", in PHAN D. (ed.), *Ontologies et modélisation par SMA en SHS*, Hermes Lavoisier, 2014.

[REG 06] REGHEZZA M., Réflexions sur la vulnérabilité métropolitaine. La métropole parisienne face au risque de crue centennale, PhD thesis, University of Nanterre, Paris X, p. 363, 2006.

[REG 09] REGHEZZA M., "Géographes et gestionnaires au défi de la vulnérabilité métropolitaine. Quelques réflexions autour du cas francilien", *Annales de géographie*, no. 669, pp. 459–477, 2009.

[SAS 02] SASSEN S., "New-York plus que jamais centre du monde", *Alternatives Internationales*, pp. 6–12, 2002.

[TUR 03] TURNER B.L., KASPERSON R.E., MATSON P.A. *et al.*, "A framework for vulnerability analysis in sustainability science", *Proceedings from the National Academy of Science*, vol. 100, no. 14, pp. 8074–8079, 2003.

[VAL 05] VALE L.V., CAMPANELLA T.J. (eds), *The Resilient City. How Modern Cities Recover from Disaster*, OUP USA, p. 3, 2005.

[VAN 01] VAN DER LEEUW S., ASCHAN-LEYGONIE S., A long-term perspective on resilience in social-natural systems, Working papers of the Santa Fe Institute, no. 01-08-042, p. 32, 2001.

[WAL 06] WALKER B.H. *et al.*, "A handful of heuristics and some propositions for understanding resilience in socio-ecological systems", vol. 11, no. 1, 2006.

[WIK 15] WIKIPEDIA, *Cyclone Lothar and Martin*, available at: www.wikipedia.org/wiki/cyclone-Lothar-and-martin/, 2015.

3

Resilience: A Question of Scale

Beyond issues to do with the defining of terms, the scientific and operational use of the concept of resilience is faced with several dilemmas:

– resilience of whom or of what?

– short- or long-term resilience?

– "top-down" or "bottom-up" resilience?

Knowing whether or not resilience is occurring always depends on the scale used: should we look at the scale of individuals, groups, territories (and if so, which ones?)? In terms of systems, how do these evolve after a disaster? How about the subsystems, metasystems, each governed by its own timescale? Is the resilience taking place in the long, medium or short term? As a result, characterizing resilience involves identifying the relevant spatial and temporal scales – which refer to the level of observation, or rather of construction, of this social object. The reference to spatial and temporal scales is, moreover, omnipresent when resilience is discussed. Several authors agree on the fact that "the degree of resilience is dependent on the pairing of spatial scales and of temporal rhythms [DAU 07]". The question of spatial and temporal scales seems to be a key one, as much when it comes to thinking about resilience as putting it into practice.

Chapter written by Géraldine DJAMENT-TRAN.

3.1. Resilience as a scalar problem

Several disciplinary and infra-disciplinary fields have approached the question of resilience from different spatial and temporal perspectives, without the choice of these scales becoming a source of debate or change of perspective. However, the scalar dimension is fundamental in characterizing resilience.

3.1.1. *Resilience and temporal scales*

Referring to resilience requires a definition for the reference state against which it can be evaluated: is it the state that immediately precedes the disturbance, or is it an older state? The idea of an original state of equilibrium is itself greatly criticized. Aschan-Leygonie has said that resilience, inasmuch as it refers to the dynamic capacity of systems to reproduce themselves, does not imply "changeless continuity" [ASC 98].

Furthermore, if we look downstream of the crisis, what is the relevant amount of time needed to evaluate resilience? A few months? A few years? or more? For example, one year after the earthquake of January 12, 2010, which resulted in 250,000 deaths and destroyed 80% of all housing, Port-au-Prince emerged from a state of emergency, as illustrated by the presidential elections, which were carried out without incident. However, despite this, the situation has not returned to normal: less than half of the 12 ministries have been rebuilt, and there is little sign of the refugee camps set up around the periphery going anywhere [COM 10]. To what extent can we talk here of resilience? While a certain amount of response time is indeed often necessary for a system to react to a disturbance, *a fortiori* to a disaster, the prolonged absence of a return to the normal situation transforms the crisis into a major and irreversible discontinuity. In this case, we should no longer talk of resilience, but rather of a chaotic situation or a bifurcation [BES 10].

From a different perspective, a system can appear to respond effectively to a crisis in the short term, while the adaptive solutions put in place can prove to be deleterious and counter-productive in the long term. In his works on the Mount Merapi [DE 12] volcano, De Bélizal showed how populations living on the slopes of the volcano, known for its pyroclastic eruptions, have turned the risk into a resource. Merapi's eruptions deposit large amounts of pyroclastic material, which interact with the monsoon rains resulting in destructive and sometimes fatal lahars. The main activity of these lands is agriculture, but, following the 2010 eruption, the farmers whose fields had been destroyed, as well as drivers and others affected, were able to reconvert themselves in a few weeks into the extraction of volcanic materials, highly sought-after for their quality, and which fuel a growing construction industry. In the short term, this extraction activity represents a form of adaptation to the crisis that allows these populations to overcome damage. However, this exploitation has some very negative effects. It increases the vulnerability of those who have to work inside the pathways of the lahars. The loss of roads, bridges and trucks, which are used in extraction, weaken the capacity to overcome the next eruption: in particular, roads are essential evacuation routes. More globally, the resource will end up being depleted while demand for it increases, as it attracts workers from further afield. These workers, who are less aware of the risks, are more vulnerable. Exploitation also causes considerable environmental damage, notably of the water table, which endangers agricultural activity. In the medium and long term, the local social and economic system, which is already fragile, becomes greatly destabilized, ultimately threatening its resilience to any new eruptions.

More globally, resilience implies the determination of thresholds, below and above which the concept is no longer relevant (or no more relevant), as a qualitative leap has been taken. Seeing as these thresholds are relative to the time considered, the discussion of timescales becomes vital in the examining of the modalities, degrees and forms of the post-crisis resilience process. The case of Haiti shows that once a certain amount of time has gone by, the lack of a return to a normal situation casts some doubt over the presence of resilience: it is less the degree of reconstruction but rather the relationship between the steps of this reconstruction and the time itself that determines whether or not we can talk of resilience.

Furthermore, this discussion allows us to classify the different *scenarii* of resilience. Thus, Vale and Campanella have identified internal temporalities, "strangely" recurring patterns, of post-disaster processes. As a result, a four-step model – each longer than the previous [VAL 05] – has been proposed. For these researchers, the processes of resilience can be split into four phases [VAL 05]: emergency response (1), restoration of that which can be repaired (2), reconstruction with the objective of achieving functional restoration and reconstruction for commemoration (3), improvement and development (4) [VAL 05]. This discretization of time is followed by the establishment of thresholds: for example, the end of the first phase is marked by the end of search and rescue missions; the third phase corresponds to the return of populations, etc.

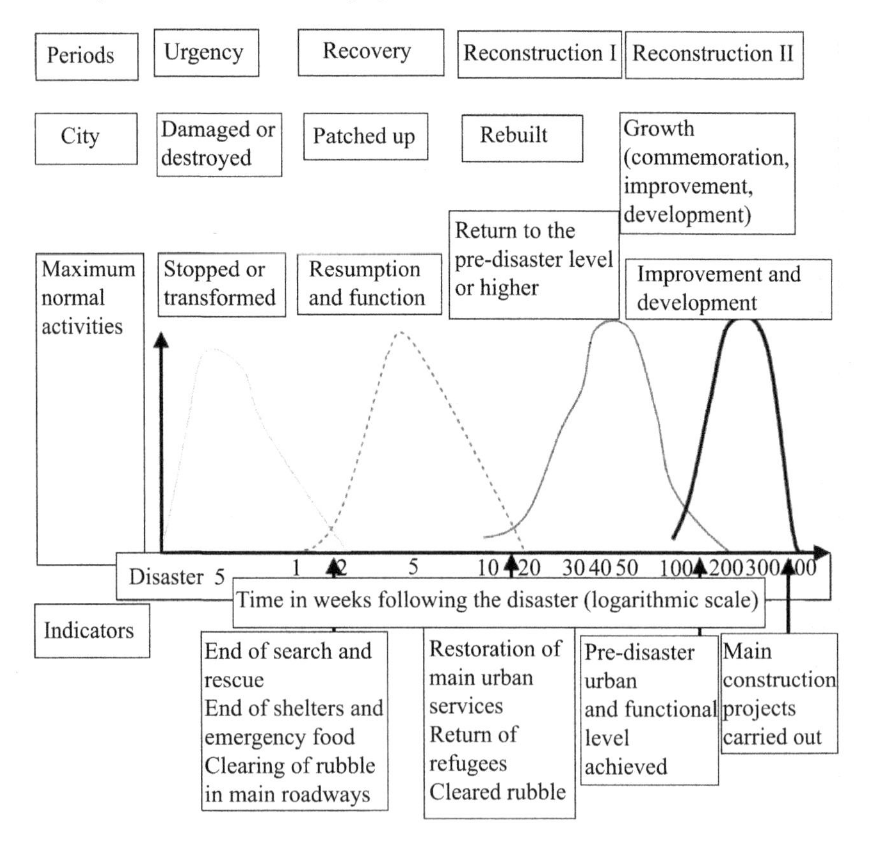

Figure 3.1. *Model of medium-term resilience (source [VAL 05], figure p. 337)*

In a certain way, discussing timescales is more important ultimately than *a posteriori* characterization of the result observed (resilient or not), or in any case, it is highly determinant in this characterization.

3.1.2. *Resilience and spatial scales*

Resilience also looks at spatial scales, covering both an issue of geographical scales as well as institutional levels.

Indeed, resilience can cover a variety of processes that occur at different scalar levels. Thus, for a city, the scale goes from that of a building to an inter-urban scale (urban resilience can refer to the resilience of a city's place within an urban city), passing through the scale of districts and of agglomerations. We can note that these spatial scales differ in terms of their temporal scales: the resilience of a building does not operate at the same rhythm as a district, which is itself governed by the different timescales of the resilience of the agglomeration. On top of the need to distinguish between different temporal scales, there is also the challenge of considering the relations between different resilience processes that take place (or do not take place) at the different spatial scales.

The epistemological difficulties associated with the geographical scale are doubly worsened by those political difficulties caused by the administrative level. Which is the institutional level (municipal, regional, national, etc.) that is the best adapted to promote resilience at a time when functional territories (whether spaces in a state of economic synergy, pools of employment, living spaces, lived spaces, etc.) no longer coincide with the limits of judicial and administrative boundaries, and where the appearance of new actors, both public and private, is dividing political territories and puts their governance into question? At a time when globalization is inducing a global rescaling [BRE 04]? At a time when risk dynamics, caused by domino effects, shift the effects of the disturbance in time and space and lead to a rapid and ubiquitous diffusion of the damage beyond the area initially affected, ignoring administrative limits?

3.1.3. *Resilience to the issue of the desynchronization of spatial and temporal scales*

Gaps between scales are a characteristic of post-disaster situations. Resilience varies depending on the spatial scales considered: some buildings can be resilient (if they are resistant or if quickly rebuilt) in a system affected by serious dysfunction and in a territory whose national and international functions are greatly affected; inversely, the resilience of a territory in most of its functions and in its image can be high, while some spaces remain in a state of crisis, or even experience bifurcation that can be the expression of "creative destruction", differential vulnerability and/or geopolitical choices. Resilience at an intra-urban level can also be unassociated with resilience at an inter-urban level. In this way, Saint-Pierre was rebuilt after the eruption of Mount Pelée in 1902, but lost out in terms of its importance in Martinique to Fort-de-France, which became the capital of the island due to its distance from the volcano.

These scalar differences are also evident at the level of temporal scales: the different components of the system considered experience recovery that is more or less sizeable. This differential resilience reveals the sociospatial structures of the territorial system concerned, as well as the political choices that are explicitly or implicitly affected. In this way, after the earthquake of 1995, reconstruction in Kobe was very rapid for economic infrastructures and fabric, such as the port, but very slow for residential areas [MEN 01]. Similarly, after Hurricane Katrina, the different districts of New Orleans presented a complex "mosaic of situations" [HER 09] with very different degrees, and, especially, forms of reconstruction, ranging from a return to normal, to a reduction in population density and to complete abandonment.

We can also note a counter-intuitive gap between the scale of the intensity of the disturbance and the temporal scale of the resilience: in New Orleans, some "*districts, greatly affected by the disaster*", such as Versailles or Mid-City, "*were able to be rebuilt quicker than districts like Hollygrove or Saint-Claude, which suffered from higher levels of flooding*" [HER 09].

We can generalize this paradox by pointing out that there is no linear relation between the intensity of the damage and the length of time needed for the recovery process. The process of resilience can be impeded by differences between the spatiotemporal scales. In New Orleans, the "prolonging of the disaster" has been due in part to significant differences between the "*timescales of planning and application*" and the "*time needed for residents to decide whether to rebuild or not*" [HER 09]: the knock-on effects of top-down and bottom-up resilience take time to set in.

Considering the multiplicity of the differences between spatiotemporal scales caused by disasters, resilience must also be seen as an issue of reconnecting and resynchronizing spatial and temporal events.

3.2. The "glocalization" of risk and scalar reconfiguration of resilience

The scalar reconfiguration that is inherent to globalization, and to what it entails, affects resilience in new terms.

3.2.1. *Globalization changes the scale of the event and of vulnerability*

Globalization changes the spatial and temporal scale of the event, and more globally of the risk [BEC 01]. We can see an increasingly clear gap for some risks (financial risk and terrorism) between the scale of the initial disturbance, often limited to a local level, and scales of vulnerability.

Due to the increasingly tight interconnections between territories, which are the foundation of globalization, highly localized events cause damage far beyond the zone directly affected; this damage can become apparent over a varying timescale. The terrorist attacks of September 11, 2001, which targeted the Twin Towers in Manhattan, had repercussions that went far beyond New York. Beyond the radical level of material destruction in the collapse of the towers, causing terrible human losses, the attacks caused repercussions in Manhattan (speeding up of the reconversion of activities [SAS 04] started before 9/11) and New York. They also caused disturbances in the economic sphere (particularly in the air travel sector), social sphere (changes in the legal and security systems in the USA and in Europe), political sphere, ideological sphere, etc. The military operations in Iraq and in Afghanistan, which were still taking place 10 years later, demonstrated this temporal and spatial gap of geopolitical consequences.

However, we can note that global-scale events come from local risks, due to the differential vulnerability of these events: thus, the global financial crisis affects some countries more than others; global climate change has consequences that are greater for some regions than for others.

There is therefore a "glocalization" [ROB 95] of risks, i.e. a telescope effect between global and local processes. Risks, and their effects, become multiscalar (they have effects at several scales), or even trans-scalar (they are diffused from the local level to the global level, or the other way around). This results in an increasing amount of immanence phenomena – a lower scale affecting a higher scale – and transcendence phenomena – a higher scale affecting a lower scale [LEV 94].

3.2.2. *Globalization leads to a re-thinking of the scales of resilience*

Globalization requires us to look into the issue of the scales of resilience in terms of the resilience of networks. By "network", we refer as much to material infrastructures as to reticular organizations, whose function is to connect individuals, locations or territories that are physically far apart.

Reticular structures present a specific type of vulnerability, which is linked to their relational potential. The interdependence that characterizes reticular organization has two negative consequences in terms of risks: it causes chain reaction disturbances (domino effects); allows for an extremely rapid diffusion of disturbances, over sometimes very large distances or surfaces. The greater the relational potential of the network, the greater are the risks of diffusion.

However, globalization is characterized by a double process: on a local scale, insertion into the process of globalization is accompanied, in large cities, or even in some particular areas of these cities, by a localized concentration of technical networks (material infrastructures) that is increasingly dense and interconnected, so as to increase the relational potential that enables exchanges of every nature with the rest of the planet, practically in real time; at a regional level, the insertion of a territory into the process of globalization results in increasingly dense relational systems, between locations and territories that can be very far from each other.

This double process creates brand new sources of vulnerability. For example,

> *The destruction of the World Trade Center towers (...) took out both a major subway transportation node in the basement and a concentration of wireless telecommunication nodes on the roof*, [MIT 05].

The terrorist attacks on a technical network in the South of Manhattan had effects that went far beyond the network, as the latter was "*a key node in the global financial network, supported by an astonishing concentration of telecommunications infrastructure*" [MIT 05]. The attacks also had consequences that were not only symbolic, but also functional over the entirety of the town, the country, and indirectly the entire world.

However, this contraction of space-time and vulnerability "*dramatically compresses the time scale for measuring urban resilience*" [MIT 05]. Indeed, the rapid return to function of transport and telecommunications networks (material infrastructures) becomes crucial: first to ensure that rescue operations are well organized and not only to help the mobilization of aid from regions surrounding and/or connected to the disturbed area, but also to allow the "normal" function of the socioterritorial system, which implies re-establishing the insertion of the system affected into the surrounding system or systems. The different phases of a return to normal mobilize different spatial and temporal scales.

The changes in the spatial scale of resilience are followed as a result by a change of temporal scale. To accommodate this, risk management needs to adapt by making predictions over greater periods of time.

3.2.3. *Who controls resilience? How to find the correct management scale*

Faced with this scalar complexification of the event and of vulnerability, the application of resilience suffers from political and administrative fragmentation, and more globally from a lack of suitability between the scales of risk management and scales which involve resilience.

Crisis management policies often collide with overlaps between the different competencies of various institutional levels.

Thus, Paris faced the risk of flooding,

> *Designing crisis management on the scale of a city is a major challenge. (...) Managing the risk of flooding theoretically needs to be done on the scale of the region of Île-de-France (...) but without heavily involving the region as an actor* [REG 06].

On top of the gap between infra-metropolitan management scales (building, district, commune and department) and the metropolitan scale, there are also those problems linked to the intervention of supra-urban levels in the resilience process of a town (national, European and global level).

Resilience can also be stimulated by actions at higher levels, but only if local knock-on effects take place. In New Orleans, federal aid economy weakened services and urban systems and proved to be a "*source of disturbance on a local scale [HER 09]*" in the medium term.

However, measurements are made to attenuate these gaps between scales. Notably, there is a search for an "optimal governing territory", or, more simply, for a relevant level of territorial government that goes beyond classical administrative limits to consider the functional coherence of the spaces concerned.

At the same time, while local and global interdependences can constitute a threat, they also represent a new opportunity to overcome disasters. Thus, the resilience process can mobilize resources from the global metropolitan network to overcome a disaster:

> *Traditionally, there was safety in numbers and in surrounding walls. Now, urban security and resilience are grounded in patterns of connectivity. And defensive rings have fragmented and recombined* [VAL 05].

At a time of general globalization, resilience is increasingly present in the positioning of globalized networks, which depends not only on the local processes of reacting to disasters, but also (or even, especially) on the interaction between the local reaction and overall dynamics. We can then imagine two *scenarii* of post-disaster reactions.

The first *scenario* consists of a weakening of resilience, caused by the change of scale: if the logic of international competition comes away victorious, the territory affected by the disaster will have some trouble ensuring resilience, i.e. maintaining its ranking, attractiveness and dynamism on a global scale.

Inversely, a second *scenario* consists of an increase of resilience caused by the change of scale: if the territorial systems which encompass the territory are mobilized, they can overcome the disaster more rapidly. Resilience is then less represented at a local scale or on a global scale, but rather in the relationship between these two. The degree of insertion into globalization and the networks associated with it then becomes a factor of resilience that helps to overcome the event.

This mechanism is not new: historically, most of the towns targeted by disasters enjoy the support of their hinterland for reconstruction, with the help of human, financial and material capital. The collapse of certain cities reveals both the sheer volume of the destruction, as well as the structural weaknesses of their hinterland. Globalization adds an extra scale, but one whose mobilization is selective. In this way, it creates new inequalities when hit by a disaster: the territories that are the best connected to the world-system, at different scales, are likely to benefit from "glocal resilience", more quickly than peripheral and different territories.

In these conditions, globalization transforms resilience into a process that is multiscalar, trans-scalar and hyperscalar at the same time. Multiscalar, in the sense that resilience not only involves local buildings and functions, but also regional, national and international functions. Trans-scalar in the sense that, due to the fact that territories connect different geographical levels, resilience involve re-connecting different scales. This aspect provides transport and communication networks within the system, a crucial role in the process of crisis management and reconstruction. Hyperscalar, as it relies on the "ability of territories to mobilize" "multilocalized resources" as part of a system of relations with multiple reaches [HAL 09]: the resilience process calls upon the complementarities of territories, as well as the interdependences within the system.

3.3. Changing scales to explain resilience

These changes demonstrate the need to change scales to analyze and explain resilience.

3.3.1. *From individual resilience to the resilience of territories*

A first change of scale consists of moving the focus from resilient individuals to the resilience of territories. Indeed the idea of contemporary resilience, popularized by international organizations, does favor the micro scale. Thus, since the end of the 1990s, the UN has developed a global strategy for resilience, the *International Strategy for Disaster Reduction*, through which the Hyogo framework of "*building the resilience of nations and Communities to disasters*" was adopted in 2007, and through which a campaign around resilient towns was started in 2010 [REV 09]. Using terminology very close to the rhetoric of *new public management* or of personal development techniques used by the big evangelical Non-government organizations (NGOs), the resilience of individuals is applauded, while the concept of secure and resilient communities is put forward and presented as a strategic element [REV 11]. Images of "resilient" individuals treated as heroes (as individuals who adapt themselves and rebuild) have gradually replaced those of vulnerable victims over the last dozen years. Notions of resilience that favor the scale of the town or city are also developed by local agents, who personify and glorify the "resilient" town.

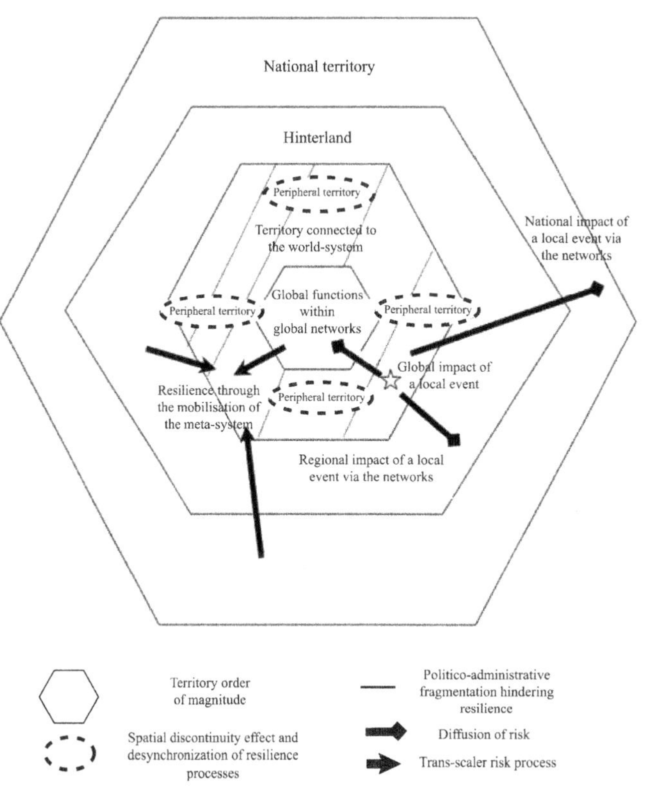

Figure 3.2. *The globalization of vulnerability and resilience (inspired by [BAU 06], Figure 6.2, "The town in globalization: a new organization")*

These scalar choices are actually important political issues: the moralization of resilience removes the initial critical weight of the concept of vulnerability, as well as the questions regarding sociospatial dysfunction revealed by the disaster, while urban myths on resilience are products of urban marketing and aim to legitimize the power of those charge and/or competition. Neo-liberal discourse concerning resilience forms a system with new scales of risk governance: the general tendency for the weakening of the state level in favor of the lower and higher levels can also be seen in the domain of resilience.

The critical reading of this rhetoric requires analyzing resilience not as a property of a level, but as a capacity to make the different scales interact with each other, to mobilize the lower and higher levels to facilitate the resilience process. For example, individual resilience appears to be conditioned by the capacity to pass from the individual scale to the scale of social networks. Thus, in New Orleans,

> *Three elements seem to favor certain individuals in their effort for reconstruction: their level of education, their competences in the key areas – whether technical or administrative, the breadth of their familial and community networks* [HER 09].

Furthermore, the resilience of a town can be explained in part by the place it occupies in the system of towns, which itself represents "*perhaps the most resilient form of all the inventions in human history [ARC 98a]*". The disturbances suffered by a town do threaten the urban network, which, by restoring its connections that link it to the town, ensures its own resilience. The resilience of a town also mobilizes its own system of relations, both close (towns and hinterland) and far (national and international networks), which can temporarily ensure the ongoing of its functions (temporary shelter of populations or of services affected by the disaster), and help recovery.

However, not all towns are able to benefit from this form of "top-down" resilience. Particularly, the crossroads situation and the hierarchical level of a town represent two major factors of urban resilience in the long term, as shown by the statistical tests carried out on an archeological database from the French South-Western urban system, "from *oppida* to metropolises [ARC 98]". Urban resilience, here measured with respect to the criterion of the sustained occupation of a site, depends on "*mechanisms of geographical selection that explain the success or decline of local subsystems*" [ARC 98b].

3.3.2. *Historicizing resilience: working with temporal scales*

Faced with a dominating approach of resilience in the short and medium terms, it would seem useful to think of resilience in the long term and situate it in an historical context, so as to improve the understanding and management of post-disaster situations.

Historicizing resilience by connecting temporal scales together first of all allows us to better understand types of reactions to previous disasters, which are partly responsible for reactions to current disasters. The diachronic comparison between different periods of crisis helps us to better grasp the effects of replay and/or learning. Indeed, considering the recurrence of disturbances in the history of towns and cities, resilience is partly determined by reactions to previous crises.

Historicizing resilience also allows us to retrace the creation of sociospatial structures, whose vulnerability can be made apparent by the disaster long after its occurrence, and thus to understand the structural dysfunction. For example, New Orleans was experiencing a state of decline before Hurricane Katrina. The urban crisis that it is undergoing currently cannot be attributed solely to the hurricane [HER 10]. We can note that such an approach implies not seeing resilience as a return to the "normal" state, the norm here being quite possibly shifted by the process of resilience itself, inasmuch as the crisis represents the occurrence of a form of geopolitical sorting, in the etymological sense of the term.

Historicizing resilience also means analyzing reactions to past crises. The anachronistic application of resilience to societies that do not use it allows us to compare the reactions of these different societies when they are faced with disasters, and thus to put into perspective the benefits, as well as the theoretical and operational limits of the current paradigm of resilience.

This paradigm fits into the crisis of the social aspect of time and into the contemporary order of historicity [HAR 03]. Thus, some towns have developed discourses of resilience before the use of the term, as seen in the paraphrase of the "Eternal City", used to refer to Rome since the 1st Century AD, or the motto of Paris *Fluctuat nec mergitur*, and strategies for recovering from disturbances, whether they are "natural", economic or geopolitical. For example, Papal Rome established a scenario for urban renovation based on the mobilization of higher spatial and temporal scales: the resetting of its international religious network and the use of long time periods, with reference to its heritage, have allowed it to overcome the polymorphous crisis of the start of the modern era, marked by the discovery of the New World, the Reformation and the Italian Wars.

This change of temporal scale in dealing with resilience implies involving works by historians [HAR 05] and geohistorians [MUS 02], who do not usually use the term "resilience", but do question the continuity of urban trajectories.

Historicizing resilience can then mean looking at the processes of resilience not in the short and medium term, but rather in the long term. The change of temporal scale moves the issue of resilience to "perennial urban trajectory" [VAL 09], or even, from a geopolitical perspective, to spatial reproduction [DJA 05], and approaches the contemporary question of durability [ARC 98]. "Perennial urban trajectory" or sustainability in the long term leads us to observe, for example, how a town reconstructs itself, from an urbanistic and/or symbolic perspective.

Historicizing resilience means finally systematically studying the scales and the temporal structures of the resilience process, using tools developed in different fields. Spatialist geography already uses transfers from the physical sciences in approaching resilience in the context of the self-organization paradigm and of the theory of dynamic systems [PUM 89, LEP 93]. Transfers from the fields of history and historiography could also be developed [POM 84, LED 99].

Resilience must be looked at from the perspective of the issue of connecting the field of experience with those expectations shattered by the disaster [KOS 79], leading us to question the shifts of timescales [LEP 93, ASC 98]. The resilience process seeks to react to events that take place at variable spatial and temporal scales, from localized to global disturbances, from very punctual disturbances (the few minutes of an earthquake) to long disturbances (economic crisis and war), or recurrent disturbances (floods, etc). For example, resilience to "natural" risks must face a telescope effect between two temporal scales[1]: "nature's" time and society's time. The different timescales of resilience itself (time to establish the vulnerable and/or resilient system, time of the disaster, response time to the disturbance, different depending on the scales considered within the system, time for the feedback after a disaster, etc.) as well as the different *scenarii* of resilience, which reveal sociospatial structures and geopolitical issues, remain paths of interdisciplinary research that need to be furthered.

3.3.3. *Scalar interactions, fundamental elements of the resilience process*

These epistemological shifts aim, *in fine*, to underline the role of relations between scales as fundamental elements of resilience.

Resilience as a reaction to geopolitical or geoeconomic disturbances frequently represents a response to change of spatial scales, to the point where we can talk of a scenario of resilience by a change of scale. Thus, Rome, the "Eternal City", which built a spatial Papal system that associates a regional State, the States of the Church, with an area of religious influence on the scale of Latin Christianity, is

1 Hazard could be defined as coincidence of two independent causal series [BAD 88]. In this case, the series interfer before the catastrophe.

found to be confronted at the beginning of the modern era with the creation of the world-system. Roman resilience involves the reuse of a global-scale religious network (change of spatial scale), but also the legitimization afforded by the great length of its ancient heritage (change of temporal scale) [DJA 11]. In the case of capitals, "*Local resilience is linked to national renewal*" [VAL 05] as a local disaster harms the reputation and the operation of the entire nation, "*and in the case of a superpower like the United States, the chain of meaning extends to the global sphere as well*" [VAL 05].

In the same way, resilience often undergoes a change of temporal scale. Individual and collective memory plays an important role in modulating the reaction to the disturbance [ASC 98]. The memory of the disaster [PIG 05], sometimes strengthened by certain politics of heritage making, is then decisive for the reconstruction capital, a concept used analogously with the concept of "immigrant capital" developed by Alejandro Portes [POR 98] and defined as:

> *The sum of material, social and cultural resources accumulated by a group to make its own reconstruction process more effective* [HER 09].

Taking into account the connections between temporal scales in planning policies can help to develop a form of proactive resilience, whether this is the "voluntary conservation of traumatic urban ruins with the goal of attenuating risk", like the ruins of Santa Maria degli Angeli in Gémona after the earthquake of 1976 [LE 10], or factoring into the planning time which is needed to obtain "disaster feedback" [ASC 98].

Resilience on a given scale can also feed off transformations, or even bifurcations from another scale. Thus, the symbolic resilience of the "Eternal City" over a long time period, in the medium term represents a lever for promoting a functional form of resilience after a geopolitical disturbance such as the Reformation. The resilience of transport networks, which maintains dialectical relations with Roman resilience, also provides information on interactions between the scales. While the "*itinerary (the flow a regional level) presents a large amount of sustainability over time* (…), *the traces rarely maintain their function as a great journey from the Antiquity to today*" and the "*model itself belongs to a relatively short timescale*": the local changes of traces maintain the itinerary in its domain of attraction;

> *This interaction between local flow, trace and model ensure a level of equilibrium between the local and the regional, which is a factor of resilience, with each scale constituting a factor of the conservation of the other* [ROB 06].

From this point of view, the gaps between scales represent a resource for the process of resilience.

3.4. Conclusion

Ultimately, resilience is characterized by its remarkable scalar polysemy (being multileveled and polytemporal). Like bifurcation, resilience appears as a scalar operator: the processes of resilience redefine the relations between scales.

Opening the "black box" of the scales of resilience leads to a lesser focus on the relevant level (or levels) and more on the political choices that are inherent to the scalar devices of resilience, as well as on the relations and interactions between the scales that make up resilience. The analytical approach which describes the different process of recovery post-disaster on different scales needs to be replaced by a synthetic approach that analyzes resilience as process that reconnects spatial and temporal scales, revealing the (dys)function of territorial systems.

3.5. Bibliography

[ARC 98] ARCHAEOMEDES, *Des oppida aux métropoles*, Paris, Anthropos, 1998.

[ASC 98] ASCHAN-LEYGONIE C., Résilience d'un système spatial: l'exemple du Comtat. Une étude comparative de deux périodes de crises au XIXe et au XXe siècles, PhD thesis, University of Paris 1, 1998.

[BAU 06] BAUDELLE G., "Villes mondiales, villes globales et city regions: trois approches de la mondialisation urbaine", in CARROUE L., *La mondialisation*, Paris, SEDES/CNED, p. 246, 2006.

[BAD 88] BADIOU A., *L'être et l'événement*, Paris, Seuil, 1988.

[BEC 01] BECK U., *La société du risque. Sur la voie d'une autre modernité* Paris, Aubier 2001.

[BES 10] BESSIN M., BIDART C., GROSSETTI M. (eds), *Bifurcations: les sciences sociales face aux ruptures et à l'événement*, Paris, Éd. la Découverte, 2010.

[BRE 04] BRENNER N., *New state spaces urban governance and the rescaling of statehood*, Oxford University Press, New York, 2004.

[COM 10] COMFORT L.K., BOIN A., DEMCHAK C., *Designing Resilience. Preparing for Extreme Events*, Pittsburgh, University of Pittsburgh Press, 2010.

[DAU 07] DAUPHINÉ A., PROVITOLO D., "La résilience: un concept pour la gestion des risqué", *Annales de Géographie*, vol. 2, no. 254, p. 117, 2007.

[DE 12] DE BÉLIZAL E., Les corridors de lahars du Merapi (Java, Indonésie): des espaces entre risque et ressource: contribution à la géographie des risques du volcan Merapi, PhD Thesis, University of Paris 1 Panthéon-Sorbonne, 2012.

[DJA 05] DJAMENT D., La reproduction de la centralité romaine. De la "Ville Eternelle" à la capitale de l'Italie. Essai de géohistoire urbaine, PhD Thesis, University of Paris 7 Denis Diderot, 2005.

[DJA 11] DJAMENT D., *Rome éternelle. Les métamorphoses de la capitale*, Paris, Belin, 2011.

[HAL 09] HALBERT L., *L'avantage métropolitain*, PUF, Paris, p. 86 and 143, 2009.

[HAR 03] HARTOG F., *Régimes d'historicité. Présentisme et expériences du temps*, Paris, Seuil, 2003.

[HAR 05] HARTER H., VIDAL L., *Mazagão, la ville qui traversa l'Atlantique: du Maroc à l'Amazonie*, Paris, Aubier, pp. 1769–1783, 2005.

[HER 09] HERNANDEZ J., "The Long Way Home: une catastrophe qui se prolonge à La Nouvelle-Orléans, trois ans après le passage de l' ouragan Katrina", *L'Espace Géographique*, vol. 38, no. 2, p. 124–138, 2009.

[HER 10] HERNANDEZ J., ReNew Orleans? Résilience urbaine, mobilisation civique et création d'un "capital de reconstruction" à la Nouvelle Orléans après Katrina, PhD Thesis, University of Paris-Ouest Nanterre-La Défense, 2010.

[KOS 79] KOSELLECK R., *Vergangene Zukunft. Zur Semantik geschichtlicher Zeiten*, frankfurt, Suhrkamp, p. 388, 1979.

[LE 10] LE BLANC A., "La conservation des ruines traumatiques, un marqueur ambigu de l'histoire urbaine", *L'Espace Géographique*, vol. 39, p. 261, 2010.

[LED 99] LEDUC J., *Le temps des historiens*, Seuil, Paris, 1999.

[LEP 93] LEPETIT B., PUMAIN D., *Temporalités urbaines*, Paris, Anthropos, 1993.

[LEV 94] LÉVY J., *L'espace légitime: sur la dimension géographique de la fonction politique*, Presses de la Fondation nationale des Sciences Politiques, Paris, p. 442, 1994.

[MEN 01] MENONI S., "Chains of damages and failures in a metropolitan environment: some observations on the Kobe earthquake in 1995", *Journal of Hazardous Materials*, vol. 86, pp. 101–119, 2001.

[MIT 05] MITCHELL W.J., TOWSEND A.M., "Cyborg agonistes: disaster and reconstruction in the digital electronic area", in VALE L.J., CAMPANELLA T.J. (eds), *The Resilient City. How Modern Cities Recover From Disaster*, New York, Oxford University Press, 2005.

[MUS 02] MUSSET A., *Villes nomades du Nouveau Monde*, Paris, Editions de l'EHESS, 2002.

[PIG 05] PIGEON P., *Géographie critique des risques*, Paris, Economica, 2005.

[POM 84] POMIAN K., *L'ordre du temps*, Gallimard NRF, Paris, 1984.

[POR 98] PORTES, A., "Social Capital: Its Origins and Applications in Modern Sociology", *Annual Review of Sociology*, vol. 24, pp. 1–24, 1998.

[PUM 89] PUMAIN D., SANDERS L., SAINT-JULIEN T., *Villes et auto-organisation*, Économica, Paris, 1989.

[REG 06] REGHEZZA M., Réflexions sur la vulnérabilité métropolitaine. La métropole parisienne face au risque de crue centennale, PhD Thesis, University of Paris X-Nanterre, 2006.

[REV 09] REVET S., "Les organisations internationales et la gestion des risques et des catastrophes 'naturels'", *Les Études du Ceri*, no. 157, CERI, Sciences Po, Paris, September 2009.

[REV 11] REVET S., Injonctions contradictoires. La gestion internationale des catastrophes naturelles: entre vulnérabilité et résilience, Séminaire Résilience Urbaine, ENS Ulm, February 2011, available at http://www.geographie.ens.fr/Compte-rendus-de-seances-2010-2011.html

[ROB 95] ROBERTSON R., "Glocalization: time-space and homogeneity-heterogeneity", in FEATHERSTONE M., LASH S., ROBERTSON S. (eds), *Global Modernities*, SAGE, London, pp. 25–44, 1995.

[ROB 06] ROBERT S., "Résilience des réseaux routiers: l'exemple du Val-d'Oise", *Bulletin AGER*, no. 15, pp. 8–14, 2006.

[SAS 04] SASSEN S., "New York reste la capitale du monde", *Alternatives internationales*, pp. 6–11, 2004.

[VAL 05] VALE L., CAMPANELLA T. (eds), *The Resilient City. How Modern Cities Recover From Disaster*, New York, Oxford University Press, 2005.

[VAL 09] VALLAT C. (ed.), *Pérennité urbaine, ou la ville par-delà ses métamorphoses*, Paris, L'Harmattan, 2009.

4

Resilience: A Systemic Property

Transposing the notion of resilience of the fields of physics or ecology into geography [ASC 98] involves questioning the trajectory of durable spatial configurations in time. For a geographer, "*any localization that is even slightly permanent, and therefore observable, corresponds to the functioning of a system that boasts a certain level of stability*" [DUR 84]. In other words, how can a spatial system maintain itself despite the disturbances and the crises that affect it over time? And, if it does last, is it still the same system?

Geographers end up testing the notion of resilience in the city considered as a spatial system or an open system, which therefore has strong ties with the environment – or meta-system – all the while drawing on recent cases of global cities that have been the victim of a disaster. What can the systemic approach add to the post-disaster resilience of an urban organism, or even of a geographical space in the larger sense of a system?

After some discussion on the different criteria that allow us to qualify a spatial system as being resilient, we will look at conditions in which we can transpose the notion of resilience to city-systems, among others, before analyzing the role of their spatial configuration in the maintenance of the qualitative structure, and therefore of their resilience after a crisis has occurred.

Chapter written by Céline PIERDET.

4.1. Resilience and systemic analysis

4.1.1. *Defining a spatial system*

There are several accepted definitions for the notion of a spatial system, which are developed in the reference work published under the direction of Auriac and Brunet [AUR 86]. For example, Auriac [AUR 83] showed how the Languedoc wine region formed as a system to explain its permanent nature. Moreover, he does not talk of "spatial systems" but refers rather to a "spatialized system". The durability of human establishments – medium-sized cities, wine-producing cooperatives, farms – and interactions that they maintain due to their closeness have enabled the persistence of Languedoc vineyards through various crises. As a result, it can be considered to be a system due to its durable presence in space. Inversely, the geographers Baudelle and Pinchemel suggest studying "*not the method of 'spatialization' of economic and social systems but rather the identification of spatial systems and how to identify their finality*", such as they are inscribed on the surface of the Earth by different societies. According to them, in the book by Auriac, "*it is the economic and social system that produces a space to last [...] such a system remains faceless, without reality: one would seek this vineyard space in vain [...]*" [BAU 86]. These two geographers consider that "*spatial systems exist as they are and not merely as spatial translations of an economic system*". In other words, "the space is a system", it is "autonomous" and is the result of anthropic actions which include actions of "identification and of denomination", of "dimensioning", "polarization", etc. which produce "*forms that are written in the ground, each family of forms with its own unique structure: paving, network, sowings*" [BAU 86].

Pinchemel insists again on the existence of interactions between spatial elements, both functional and morphological, from which the "geonomic" spatial system originates [PIN 97]. It goes back to the overlapping of scales, and therefore of spatial systems and subsystems, the association of the "three constituents of space" that are surfaces, lines and centers and to which any element on the surface of the planet can be assimilated, no matter its scale. The great number of possible associations between these elementary forms, as much in two dimensions as in three, as well as the various overlaps of scale, creates an infinitely diverse amount of spatial systems. They are the result of a human desire to organize space, with each society producing its own models. Marked by certain highly permanent aspects – fragmentation, containment, town centers, administrative limits, etc. – they are nonetheless continuously subject to change. The modification of an elementary component has repercussions on the system as a whole.

The spatial system of mining [BAU 94] is an example of a system that revolves around one functional objective. For Baudelle and Pinchemel, this type of system usually overlies previous spatial systems, often agricultural in nature, which never quite disappear. The appearance of this new system, whose finality is the production of coal at the lowest cost possible, is the result of anthropic actions: "identification" with the search for coal fields and the "denomination" of different concessions, the "dimensioning" of these concessions, islets, roads, etc. Several poles – wells, factories and cities – appear as a result. The two geographers explain that "*the mining system makes up a coherent system: co-spatiality and interdependence make it a spatial system*". From one mining system to another, the expansion of the spatial system allows for an increase in production. It is a lack of labor, accommodation, land, the exhaustion of coal deposits that regulate the growth of the system and help avoid its entropy. The mining system is therefore considered to be an open system, which interacts with its environment.

The pioneering works by Le Berre on the notion of "territory" help link together space and social groups during the production of a spatial system. She proposes defining this system as:

> *An organized entity that evolves in an environment in function of the interactions between a social group and its territory. (...) the social group, actor of the spatial system, produces the territory, maintains itself and reproduces itself within it; the territory, which is the group's living space, provides the conditions needed for this maintenance and this reproduction: it is therefore also an actor of the spatial system* [LE 92].

We can see here how the appropriation of space by a social group results in a territory, and how, by retroaction, we would witness the appearance, and then the expansion, of a spatial system.

4.1.2. *Criteria for a resilient spatial system*

According to Holling [HOL 86], two conditions must be met for resilience to be adapted to systemic entities other than ecosystems. First, the system must not be static, but rather changing. Second, several meta-stabilities – or unstable, dynamic equilibriums – must be possible around the same attractor, the same trajectory, which is indeed the case for spatial systems. This new definition of resilience reflects the possibility, for a system, to experience several metastable states within the same domain of attraction. This goes against the traditional definition according to which a return of the system to a situation present before the crisis is the only state possible for resilient behavior. Therefore, a system can be far from equilibrium while still absorbing the disturbance, functioning and remaining in the same domain of attraction. If this is not the case, it changes qualitative structure and bifurcates. However, the fluctuations measured by the trajectory of a system vary depending on the timescale considered [DUR 99]. Each temporal scale corresponds to different and overlapping spatial scales: the spatial system and its components absorb a disturbance and change depending on differentiated temporalities.

In the mid-1990s, geographer C. Aschan-Leygonie carried out part of her doctoral research at the International Ecotechnology Research Center in Cransfield, UK, as part of the European Archaeomedes project. This research center contributes to the transferring of concepts in hard sciences toward social sciences. Aschan-Leygonie worked on transferring the notion of resilience as present in ecology toward geology, by applying it to the region of Comtat-Venaissin, whose agriculture was affected by two large crises in the 19th and 20th Centuries [ASC 98]. The methodology relies on the qualitative formalization of a system from a critical review of the literature, as well as on the consideration of statistical data, from a diachronic perspective. She describes the Comtat system as:

> *A structure based on agriculture production that is majorly speculative, small family-run businesses with direct return, drawing on a sizable proportion of seasonal workers, and a dense network of markets and small towns where food products can be sold. The interactions between these different elements have led to the maintenance and the reproduction of this structure over a relatively long time period* [ASC 99].

How did the spatial system of Comtat become resilient so as to maintain itself over time beyond the crises? For the second half of the 19th Century, Aschan-Leygonie identified the "*arrival of a set of brutal disturbances that added themselves to pre-existing cyclical disturbances*" [ASC 99]. Viticulture, for example, was hit by phylloxera, while the arrival of silk in Japan led to a drop in the price of raw silk and caused the ruin of silkworm breeders. The area occupied by grapevine decreased in 1870, while the madder disappeared entirely. The effects of agricultural crises combined with the development of rail travel affected all production in the region. However, these external disturbances actually reinforced the interactions between networks of towns, irrigation and transport links, and helped to maintain the spatial system. Thus, the Comtat system absorbed these disturbances in the 19th Century, while in the 20th Century, as part of a context of the growth of the European Union, of competition and of the globalization of commerce, the fruit and vegetable sector struggled to adapt. These slower disturbances relativized the notion of early growth produce from Comtat at the scale of the European Union. Greenhouse vegetable cultivation became prominent and large surfaces challenged the traditional circuits of distribution. As a result, the system entered a state of crisis.

Finally, C. Aschan-Leygonie highlights:

> *The size and the speed of the introduction of a disturbance into a system (...) to understand its dynamics (...); the characteristics of the functioning of a system at the moment where the disturbance affects it are also elements that can explain its capacity to maintain itself during and after the crisis* [ASC 99].

Therefore, a system can only be considered resilient *a posteriori* and different properties are necessary for it to develop such a behavior. First of all, its lack of stability allows it to absorb a disturbance. It can be defined as an "*indicator of the size of the domain of attraction around an equilibrium in which the system can evolve, without changing qualitative structure*" [HOL 73]. Next, the components must exhibit a large amount of diversity. For example, for ecosystems, Holling remarks that some modern and intensive agricultural systems are stable, but a lot less resilient than more extensive traditional systems. The small amount of diversity seen in the cultures results in high sensitivity to illnesses, attacks from parasites, as in the case of single-species stands of trees in forests. Finally, it must exhibit a certain degree of adaptability. Faced with a disturbance, a system whose actors develop pioneering behavior will find it easier to adopt innovations, whether they are technical, institutional, etc. Furthermore, these actors facilitate the diffusion of information within the system. In this way, its capacity of adaptation to a disturbance is linked to the quality and the quantity of the interactions maintained between its components, as well as the behavior of its actors faced with the crisis.

4.2. The case of the city, a complex sociosystem

4.2.1. *The synthetic approach to systemic analysis*

Since the start of the 1970s, in a context of a great amount of uncertainty over the future of cities due to rapid mutations in the behavior of urban actors[1], it has become normal to assimilate cities to systems in which each of these transformations interacts with all of the structure over its function and its evolution over time. The systemic approach involves considering that the components of the city-system maintain interactions in the time that can be described, from differential equation systems for most of them, so as to take into consideration this historical dimension, or even to anticipate

1 E.A. GUTKIND [GUT 62] talked here of the "twilight" of towns, L. MUMFORD [MUM 70] of the "decline" and P.H. CHOMBARD DE LAUWE [CHO 82] of the "end of towns", see [PUM 89].

its trajectory. After the genesis phase of the system – the systemogenesis – these interactions create positive retroactive loops – the growth phase of the system – or negative retroactive loops – the regulatory phase. In the case of a complex system like a city, which is the result of a large variety of elements and a multitude of interactions between these elements, the succession of these retroactive loops makes the trajectory nonlinear over time, and therefore is made up of phases of growth, stasis, regression, etc. Internal fluctuations or external disturbances can lead to a reorganization of the town-system. A disaster can also take place and change the qualitative structure of the system, and therefore cause it to bifurcate from one trajectory to another, or lead to its collapse. With the exception of this extreme case, adjustments are constantly being made between the interacting subsystems within the same system.

The social systems analyzed by the scientist J.W. Forrester – industrial, urban, etc. – are considered to be closed, autonomous and stable to describe their macroscopic structure and anticipate their evolution [FOR 79]. To explain their behavior, he considers "*interactive retroactive loops that develop their effects within a closed system (...)*" [PUM 89]. For J.W. Forrester, positive and negative retroactive loops, as well as the variables presented in the form of flows that enter and leave the system and in the form of stock, are at the center of the functioning of the system. It is the level reached by the variables of the stock – linked to the intensity of the flow – that is an indicator of the state of system. This vision of a city as an autonomous sociosystem has been criticized by geographers like G.P. Chapman [CHA 77], due to the many interactions that it necessarily must have with the outside.

Inversely, the scientist J. de Rosnay hypothesized in 1975 that:

> *Growth and illnesses of the city, the multitude of its functions, its daily behavior suggest that the city reacts like a living organism that communicates with an environment that it modifies indirectly and which, in turn, shapes it* [DE 75]:

Moreover, he states that:

> *An open system is permanently in touch with its environment (as a generalization we could say its ecosystem). It exchanges energy, matter, information used in the maintenance of its organization against the damage that comes with time. It rejects entropy into the environment, or 'used' energy* [DE 75].

The functioning of systems is illustrated by J. de Rosnay using hydraulic models[2], combining among other things retroactive loops, flows, valves and reservoirs. The town can be viewed as an open system, with unclear limits, and

> *The maintaining of its identity is a dynamic phenomenon that implies [...] the exporting of entropy toward the outside, but also a continuous increase of complexity by successive transformations as part of the process of self-organization* [PUM 89].

Relations with the outside, which enable this self-organization, are at the same time exchanges of materials, people, information, etc.

Finally, the modeling of interactions within a system also leads to an interpretation of the "regularities" observed in different systems [PUM 89]. For geographer R. Brunet, the observable spatial models can be reduced to around 40 "*chorèmes*", or elementary structures, analogous to attractors for spatial systems. However, the state of a system at a given moment can also be explained by taking into consideration a past that alone can help understand the trajectory that is followed and that makes it unique.

4.2.2. *The nature of interactions in a spatial system*

The cohesion of a system depends on the intensity of the interactions between its components. These interactions ensure a rapid

2 The figures in this book were created by J.W. Forrester.

speed of diffusion of information within the system, and therefore its capacity to dampen the disturbances that affect it. C. Aschan-Leygonie reminds us that spatial systems:

> *[...] are defined both vertically, in terms of the relations between the human group, its activities, the environment and the inherited spatial structures, and horizontally, by the spatial interactions between elementary geographic entities. Vertical and horizontal relations are strongly interdependent* [ASC 98].

Quantitative modeling generally seeks to predict "*changes of the structure of the system*" [PUM 89] by assuming that they are the product of changes in the interactions between the elements of the system. It links the evolution of interactions at the microscopic level, between the elementary components of the system, and their repercussions on the macroscopic structure. The models can anticipate several trajectories and situations of equilibrium, or even bifurcations of the system.

A lot of work in the human and social sciences also cover systemic analysis and the question of interactions in a qualitative manner. A spatial system appears when two spaces are differentiated and when flows, exchanges develop between them due to their complementarity [DUR 92]. As a result, a positive retroactive loop is established and reinforces their specialization, and the system experiences expansion due to exchanges between two spaces, feeding their development and their reciprocal growth. Within the spatial system of mining analyzed by G. Baudelle, this heterogeneity causes, for example, a drain of labor between the countryside and extraction sites that maintain the cohesion of the system.

How do spatial interactions maintain cohesion when faced with a disturbance? F. Auriac shows how the interactions between medium-sized cities, wine-producing cooperatives and farms are factors of the resilience of the "spatialized" Languedoc system after two crises recorded in the 19th and 20th Century. These interactions "*allow the effects of the disturbance to be shared between the multiple spatial components, decreasing in this way the intensity of its effects*"

[ASC 98]. Spatial interactions are also at the root of the diffusion of information and of "innovations" within the system.

We have also looked at the social spatial resilience of Phnom Penh, the capital of Cambodia, from the fall of the Khmer Rouge regime in 1979 [PIE 08]. From the works by J. de Rosnay, among others, we can assimilate a *hydraulic system* with the urban organism that is surrounded by dams, subdivided into catchment areas, crossed by a sewerage network that evacuates used and rain water away from the city center, and whose different components maintain strong interactions between themselves and with the environment.

Phnom Penh was practically emptied of its inhabitants on April 17, 1975 by soldiers of the Khmer Rouge and their leader, Pol Pot[3]. At the end of the dry season, the valves through which dirty water from city flowed were all open. From 1975 to 1979, the river overran the sewerage network during the annual freshet period, silting it up and rendering it useless, as the Khmer Rouge did not know how to control the hydraulic infrastructure. Only evaporation and gravitational flows were able to ensure the return of the city-system to equilibrium during the dry season, due to low population levels. This infrastructure was therefore greatly damaged and run-down when, in January 1979, Phnom Penh was taken by Vietnamese troops, while the population flocked to the capital. From 1980, faced with the grave state of the sanitary situation, "pioneers" [ASC 00] – engineers, international experts, members of non-governmental organization (NGOs) – sought to reconstitute the data destroyed under the Khmer Rouge, necessary for the appropriation of the land. Despite their modest means due to an international embargo being placed on Cambodia, punctual interventions were carried out on the network: unblocking sections of the canals, repairing faulty pumping stations, etc. At that point, the goal was to maintain the state of the network, short of being able to carry out any overall work, and to avoid letting the situation get worse.

3 The Pol Pot regime – deurbanization, elimination of the elite, etc. – was responsible for a genocide estimated at 1.7 million victims.

The recovery of political institutions, the mobilization of exogenous professional cultures associated with a punctual rehabilitation of the network, have aided the resilience of the system by giving it a faculty of adaptation to new techniques, as well as by maintaining the diversity of the components and of the interactions or levels within the city-system. Strong interactions were returned from 1991 onward between the city-system and the environment with the establishment of a new watershed and the recovery of the hydraulic backfill. Water remains an entrance, a component and an exit of the system, and still singularizes it as a *hydraulic system*. A slow process of "reterritorialization" and of reclaiming of the hydraulic system then takes place.

4.3. Maintaining the cohesion of the system to overcome the crisis

4.3.1. *A dichotomy of "kernels"/"margins"*

There are many spatial systems that have recorded times of considerable crisis, including in the contemporary period. An extreme case like that of Chernobyl is an exception, however: the fusion of reactor 4 of the nuclear reactor, on April 26, 1986, led to the evacuation and abandonment of the town by the authorities, as life had become impossible due to the high levels of radioactivity. Outside of this case, these spatial systems entered into resilience and were able to last. How can the spatial dichotomy of a city-system be a factor for durability and resilience after the occurrence of a disaster? What are its characteristics?

C. Aschan-Leygonie looks at the "*nature of limits in space. A spatial element can simultaneously be part of several geographical spaces*" [ASC 98]. By calling upon the terminology used by R. Brunet [BRU 90], she considers there to be "kernels" "nuclei" that are therefore "*defined by their high chance of belonging to a single system*" and "margins" that are "*characterized still by their belonging to a single kernel, although less clear*". The kernel of the system is characterized by its high levels of coherence, and its homogeneity. The margins are, by definition, more heterogeneous and less structured. The whole forms a system due to the vertical and horizontal interactions that enable it to maintain its coherence.

Within a spatial system, these very indured "kernels" play the role of "transmitters" of innovation, while the less indured "margin" are more permeable to innovation and act as "receptors" [BRU 90]. Therefore, innovations are diffused from the kernels toward margins within a spatial system, due to interactions maintained between the different components of the system.

4.3.2. *Disturbances with different origins*

In the trajectory of a spatial system, a disturbance plays the role of a regulator during the regular growth phase of a system, and helps to avoid its entropy and its indefinite expansion. In the opposite case, this system would then be "*characterized by high levels of homogeneity and by an extreme simplification of the structure that would risk compromising its maintenance. [...] The appearance of disturbances favors the system's renewal*" [ASC 98]. A disturbance can have an exogenous or endogenous origin, appear progressively or suddenly. An exceptional meteorological event can thus affect the system quite brutally and often unpredictably.

Thus, the terrible floods that took place in the flood plain of Menam Chao Phraya in Thailand, during the fall of 2011, are cyclical since they followed the monsoon rains that happen every year in Asia, between the months of April and November [PIE 12]. They are made worse by an important subsidence process that has been affecting the city of Bangkok since the end of the 1970s, due to the rapid upward growth of constructions and considerable pumping of the water table. The average annual waterfall in Bangkok is of 1,450 mm, but 50% of this concentrated between the months of September and October. In 2011, the provinces located between the north of the flood plain and the north of the metropolitan region of Bangkok, which includes the main industrial regions of the country, remained flooded for over two months. Several districts of the city also had to be evacuated for several days by the government at the end of the month of October, at a time when big tides in the Gulf of Thailand were stopping the draining of river water. Several dams burst and the hydraulic infrastructure – canals, storage bays, pumping stations, etc. – always increasing in size and number, proved to be insufficient. The presence of land reserves allowing the freshet water to spill over around the outskirts of Bangkok remains crucial in keeping water away from the inhabitants and economic activities, but it is threatened by the pressure of the growth of city and the expansion of urban territory.

While the events are indeed natural in their origin, the amplitude of the disaster is modulated by choices made in terms of construction planning, particularly in metropolitan areas, which are determining the presence of resilience, or its absence. Most of the time, it is "*an overlap of different types of disturbances that affect the system*" [ASC 98]. In other words, a punctual disturbance can fit in with a general change of context in which the system evolves, and therefore over longer timescales. Depending on the origin, the amplitude, the brutality and the timescales of the disturbance, it either is or is not integrated into the function of the system, and this system will either be able to maintain itself, or not. An overly brutal disturbance, which takes place during a phase of low stability of the system is very likely to change its qualitative structure and domain of attraction, and therefore to make it undergo a bifurcation.

4.3.3. *The system's response to disturbances*

C. Aschan-Leygonie remarks that:

> *Depopulation, a corollary of large crises, is a form of behavior that is increasingly adopted as a response as the system gets close to the superior limit of the population load that it can support in the context that preceded the crisis* [ASC 98].

This was verified in 2011 in Bangkok. The Thai government proceeded to the evacuation of the central districts of the capital for several days at the end of October. Considering the strong pressure caused by the floods on the big dams that protect the city, the authorities had to cause breaches themselves, and took the decision to flood some of the central districts to lower the level of the body of water and avoid an uncontrolled and sudden rupture of the main dams. As a result, the flood blocks became uninhabitable. During any major crisis, the evacuation of populations helps to reduce the network load and reduce the instability of the system as a whole, until it stabilizes around a new equilibrium.

We must make the distinction here between the temporary evacuation of a city, during a crisis like those that occurred in New Orleans in 2005 or in Bangkok in October 2011, from a near-complete evacuation of several years, like the one that took place in Phnom Penh between April 1975 and January 1979. At the time of evacuation, the population of the Cambodian capital reached a peak level estimated to be roughly 1.5 million people, which was not matched again until the census of 2004. In this case, there is a marked decrease in the interactions between some of the components of the system, due to the absence of upkeep of its technical networks, for example, and due to the multiple attacks carried out during this period. The ultimate stage of disorganization would have been reached in the case of total and definitive abandonment of the city-system leading to its collapse. This was not the case here, but the return of the population to a city left practically abandoned for several years was initially highly controlled and progressive, with the networks clearly unable to support a sudden massive load, because of the amplitude of various dysfunctions. Moreover, institutional disorganization is durable in such a context, and the return of vertical interactions between inhabitants, the various actors and components of the system is often very slow. The massive and sudden occupation of the system operating in a mode of damage makes it more unstable and can lead to it bifurcating, or changing its domain of attraction.

The resilience of a city-system having absorbed a large disturbance is both spatial and social in nature, but also "institutional" [ADG 00]. The return of vertical and horizontal interactions within the system – both in density and in intensity – depends on the prompt recovery of several vital institutions – administration of political power, intervention of "pioneers", municipal technical services, health services, markets, etc. If the diversity of the components of the system have been properly conserved, as soon as the horizontal and vertical interactions are recovered, a positive retroactive loops is formed again, ensuring the growth of the system takes place. During this new phase that leads to resilience, the "pioneers" can harness the need to replace most hydraulic infrastructure, for example by modernizing it, to increase the evacuation capacity of the network using larger canalization or more powerful pumping stations. They enable the system to adapt to the post-crisis system, increasing its capacity for resilience. This is a form of "creative destruction" in the Schumpeterian sense, which can be supported by using new actors such as bilateral or multilateral cooperation, through the support of fund providers, especially in developing counties. The city-system is then ready to receive technical and cultural innovations due to its capacity for adaptation. In the case of Phnom Penh, the modernization of hydraulic infrastructures started in the 2000s, with the successive replacement of the components of the network in the most urbanized areas of the city center, due to providers of funds – the World Band among others – and bilateral cooperation with Japan. These innovations were then diffused from the "kernel" or "transmitter", toward the "margins" still called "receivers" [BRU 90] and contribute toward the resilience of the city-system.

4.4. Conclusion

The systemic approach enables us, among other things, to apprehend the trajectory of an urban organism or a diachronic territory, and to understand that it is made up of a succession of differentiated phases that, beyond the organizational similarities of spatial systems, make them unique. Therefore, the trajectory of a city-system is not linear. Over time, a system integrates into its functioning the effects of past disturbances so as to come up with adapted responses to new crises, which are factors of resilience. Similarly, while the organization of a resilient spatial system possesses regularities, it is also characterized by the large diversity of the components which define it. The various actors of the system that follow each other enable the receipt and diffusion of the innovations within the system, as well as its adaptation to factors of instability. Paradoxically, the actors of a system that regularly records disturbances also keep a better memory of these disturbances and of the responses that must be supplied. Dynamic, in a state of unstable equilibrium, the spatial system adapts itself and enters into resilience.

4.5. Bibliography

[ADG 00] ADGER N., "Social and ecological resilience: are they related?", *Progress in Human Geography*, vol. 24, no. 3, pp. 347–364, 2000.

[ASC 98] ASCHAN-LEYGONIE C., La résilience d'un système spatial: l'exemple du Comtat. Une étude comparative de deux périodes de crises au XIXe et au XXe siècle, PhD Thesis in geography, University of Paris 1, 1998.

[ASC 99] ASCHAN-LEYGONIE C., "Temporalités et résilience d'un système – La dynamique comtadine au XIXe et au XXe siècles", *TIGR*, nos. 101–104, pp. 63–82, 1999.

[ASC 00] ASCHAN-LEYGONIE C., "Vers une analyse de la résilience des systèmes spatiaux", *L'Espace géographique*, no. 1, pp. 66–77, 2000.

[AUR 83] AURIAC F., *Système économique et espace: le vignoble languedocien*, Paris, Economica, 1983.

[AUR 86] AURIAC F., BRUNET R. (eds), *Espaces, jeux et enjeux*, Libr. Arthème Fayard, 1986.

[BAU 86] BAUDELLE G., PINCHEMEL P., "De l'analyse systémique de l'espace au système spatial en géographie", AURIAC F., BRUNET R. (eds), *Espaces, jeux et enjeux*, Libr. Arthème Fayard, pp. 85–94, 1986.

[BAU 94] BAUDELLE G., Le système spatial dans la mine: l'exemple du bassin houiller du Nord-Pas-de-Calais, PhD Thesis, University of Paris 1, 1994.

[BRU 90] BRUNET R., "Le déchiffrement du monde", *Mondes Nouveaux, Géographie universelle*, Belin/Reclus, Montpellier, vol. 1, p. 158, 1990.

[CHA 77] CHAPMAN G.P., *Human and Environmental Systems, a Geographer's Appraisal*, Academic Press, London, p. 421, 1977.

[CHO 82] CHOMBARD DE LAUWE P.H., *La fin des villes: mythe ou réalité*, Colmann-Lévy, Paris, 1982.

[DE 75] DE ROSNAY J., *Le macroscope. Vers une vision globale*, éd. du Seuil, Paris, p. 55, 1975.

[DUR 84] DURAND-DASTÈS F., "La question "Où?" et l'outillage géographique", *EspacesTemps*, nos. 26–28, p. 11, 1984.

[DUR 92] DURAND-DASTÈS F., GRATALOUP C., LEVALLOIS L., "Le rôle des flux dans l'organisation des ensembles spatiaux", *L'Information géographique*, no. 1, pp. 35–42, 1992.

[DUR 99] DURAND-DASTÈS F., "Jamais deux fois… Ou: De quelques précautions à prendre avec le temps", *TIGR*, nos. 101–104, pp. 5–23, 1999.

[FOR 79] FORRESTER J.W., *Dynamiques urbaines*, Economica, Paris, (1st edition 1969), 1979.

[GUT 62] GUTIKIND E.A., *The Twilight of Cités*, Free Press of Glencoe, 1962.

[HOL 73] HOLLING C.S., "Resilience and stability of ecological systems", *Annual Review of Ecology and Systematics*, vol. 4, pp. 1–23, 1973.

[HOL 86] HOLLING C.S., "The resilience of terrestrial ecosystems; local surprise and global change", in CLARK W.C., MUNN R.E. (eds), *Sustainable Development of the Biosphere*, Cambridge University Press, Cambridge, pp. 292–317, 1986.

[LE 92] LE BERRE M., "Territoires", in BAILLY A., FERRAS R., PUMAIN D. (eds), *Encyclopédie de Géographie*, Économica, Paris, pp. 617–638, 1992.

[MUM 70] MUMFORD L., *The myth of the Machine: Volume II The Pentagon of Power*, Harcourt Brace Javanorich, 1970.

[PIE 08] PIERDET C., Les temporalités de la relation ville–fleuve à Phnom Penh (Cambodge) – La fixation d'une capitale fluviale par la construction d'un système hydraulique (1865–2005), PhD Thesis, University of Paris I – Panthéon-Sorbonne, November 2008.

[PIE 12] PIERDET C., "La résilience comparée de Phnom Penh (Cambodge) et Bangkok (Thaïlande) face aux crises hydrauliques", DROGUE G. (ed.), "Climats et changement climatique dans les villes", *Climatologie*, Grenoble, AIC, pp. 83–108, May 2012.

[PIN 97] PINCHEMEL P., PINCHEMEL G., *La face de la Terre*, 5th edition, A. Colin, 1997.

[PUM 89] PUMAIN D., SANDERS L., SAINT-JULIEN T., *Villes et auto-organisation*, Économica, Paris, 1989

5

From the Resilience of Constructions to the Resilience of Territories: A New Framework for Thought and for Action

All territorial systems, and particularly urban systems, link together a series of components that allow them to cope with threats, hazards and risks. These elements are not only material in nature. The literature is often ambiguous on the matter, as it tends to group the building (physical component of the system) together with its function. This ambiguity is often highlighted in studies that use the concept of resilience in relation to the notion of absorption or of "*recovery*" [HER 09], reminding us that resilience relates as much to materials as to customs, representations, landscapes, living spaces, lived spaces, urban identities, etc. This is still true on several scales. However, this socioterritorial resilience relies undeniably on the resistance of constructions: commerce, as a social organization, depends on the buildings that host the businesses. For example, the activity of nearby businesses is necessary for resilience as it guarantees at least a minimal level of quality of life during a crisis, but can only be ensured if the building itself is not too damaged.

Designing a resilient territory therefore also involves creating buildings that are able to resist destructive events by choosing adequate materials and architectural techniques, even if this means

---

Chapter written by Bruno BARROCA.

organizing how the building is used so as to limit the functional impact of any possible damage.

As far as illustrating possible ways of developing resilience as a capacity for resistance goes, the town is a privileged space. Indeed, in recent years, a form of urbanism that is specific to high-risk areas has developed, which seeks to improve the resilience of constructions, to limit material damage, as much as enabling resilience in the urban system as a whole – referred to as continuity of activities, the recovery of functions, or a return to normal. This urbanism must work on two scales: first scale of the urban system, and the second scale of the elements that make it up.

5.1. The conditions of resilient planning on the scale of the territory

Any policy that seeks to manage risks by developing resilience cannot operate independently of the territory concerned [NOV 11, NOV 02]. In the context of urban spaces, this means developing an integrative vision, which does not discriminate between risk management and town management, whether this is the work of planning, urbanism or local development. It is by thinking about the future of the town and urban planning that the measures of risk prevention are defined, and *vice versa*.

5.1.1. *Project urbanism, a new reference framework*

To develop the resilience of a place, risk must be considered not as an external element, but rather as a component of this place [BEU 08]. As a result, urban planning and risk management must be brought together so as to take into account their clashes and synergies, while technical and engineering aspects must be discussed and confronted by institutional and political dimensions.

In practice, the goal of resilience is indeed defined by politics. However, its operational application, and even more so the effectiveness of the devices, assume an association between all of the actors involved in planning and in risk management.

Mixing these two logics of planning and prevention therefore implies a capacity for partnership between the managers (especially State services in France) and the community successively to:

– ensure a shared definition of the issues of the approach and ensure management in a long-term perspective;

– establish an urban diagnosis that integrates the issues of development and a review of the risks. These works form the basis of the mobilization of all the stakeholders (together with management of risks, of urban and architectural shapes, of economic hazards, of urban services and of the environment). Following the sharing and the validation of the urban issues and of the situation, shared strategies for resilient urban renewal are designed in order to cope with risks;

– identify, following the study phase of the project, alongside all the relevant stakeholders, methods of funding, defining everyone's role in the application and validating operations phasing.

This requirement for an integrated and transverse approach implies the establishment of new processes, with a change of urban practices and of the reference framework that they sit in. Notably, it calls for the end of "product or norms-based urbanism", instead replacing it with a more actively participating project urbanism.

In France, the process of project urbanism ("*urbanisme de projet*"[1]), started in 2010, following the *Grenelle Environnement*[2], would appear particularly welcome here. *This new approach* goes beyond the accumulation of rules or even the length of procedures and of conflicts. It takes into account all problems, but also the resources and potential of the territory, integrating technical, economic and human aspects, mobilizing the different actors and searching for

1 This term is not easily translated into English. "Urbanisme de projet" is a new experimental practice in France with less regulation than the usual practice.

2 The "Grenelle Environnement" is a conference bringing together the government, local authorities, trade unions, business and voluntary sectors to draw up a plan of actions of concrete measures to tackle environmental issues.

synergy between managing the building, the project and managing use. The execution of complex or ambitious projects requires the coordination and coherence of procedures and prescriptions that, these days, tend to join together without any coordination, or even contradict each other. This logic becomes an injunction against the imperatives of sustainable urban development: it is particularly relevant in places threatened by risk, as exposure to danger is an important source of ecological inequality, often encompassing socioeconomic inequalities [EME 07]. Both the objectives of resilience and of urban valorization must then be pursued simultaneously.

5.1.2. *A risk-integrating place-based project*

This approach of project urbanism requires the devices of high potential for adaptation to local specificity to enable proper integration of the risk into the urban project by mobilizing all of the stakeholders of urbanism, risk management and construction.

The "lab town" model proposed by C. Redeker for the construction of the resilient district of Mayence can provide a good example to follow. The principle relies on interaction and sharing between stakeholders, methods of evaluation and of design, and aspects of management, through operational application. This approach considers resilience by structuring the actions according to the model developed and tested by the Scottish government, called the "4 A's" method. This method takes its name from the four themes that make it up: Awareness, Avoidance, Alleviation, and Assistance.

1) *Awareness* consists of not only making sure the population is aware of the risks, but also of the ways to mitigate these risks, whether on the level of:

– those taking decisions (politicians and deciders);

– professionals (connected institutions or others);

– the public (populations, businesses, developers and insurance companies).

2) *Avoidance* consists of limiting the fragility of urban components. This also involves an easier and faster phase of reconstruction, and therefore, for example, training populations and institutions for reconstruction. Adapting buildings and infrastructures, planning the layout of urban space, are all good methods to achieve this.

3) *Alleviation* involves the containment of the hazardous event, and therefore the reduction of risk, usually by putting in place physical or technical infrastructure.

4) *Assistance* can correspond to the capacity of citizens to take action before, during and after the crisis.

A fifth axis can be added to these themes: "Competence and Strategies", which helps engage the processes mentioned above.

This general approach that links risk with planning and urbanism, leads the drivers of the project to identify the urban components that need to be adapted to obtain high levels of resilience. The case of Villeneuve-Saint-Georges[3] illustrates the integration of works of planning and of risk management. The requalifying of the town center of Villeneuve-Saint-Georges is one of the priorities declared by the public planning establishment Orly-Rungis Seine-Amont (EPAORSA). This operation of urban planning is part of an approach for sustainable urban development, oriented toward an increase in hygiene levels and of resilience.

The town center of Villeneuve-Saint-Georges is located between a hill and a linear backfill built for the railroad. This backfill forms a bank that is somewhat like a dam, protecting the city from floods coming from the Seine. Diagnostic tests show that the rainwater sanitation network is currently saturated and overrun when significant events take place. Between 1982 and 2005, Villeneuve-Saint-Georges was the object of 18 inter-ministerial orders declaring a state of natural catastrophe. Moreover, in the case of a high-volume freshet of the Seine, infrastructure in place stops the water going back up through the networks.

3 Villeneuve-Saint-Georges is a city of 10,000 inhabitants located in the Paris region 16 km southeast of Paris.

To enable the evacuation of water from the network, pumps were installed. These pumps play an essential role; however, if the Ile-de-France has to deal with the centennial flooding of the Parisian basin, the pumps' energy supply would almost certainly not be maintained.

EPAORSA (the planner) has worked with local collectivities and called upon teams of architect-urbanists who are experts in risk. One specific approach enabled the evaluation of whether resilience to flooding is increased by the planning project. This independent study from the teams of architect-urbanists relies on a group of experts in risk management who develop the approach and communicate with the management team.

Different technical solutions have been proposed to adapt the urban space to hydrological risks: changing the infrastructure, urban morphology and new buildings, transforming that which already exists. The requalifying of the town center makes sure that the soil sealing is not increased. Seeing as the current situation is not acceptable in terms of the management of rainwaters, the designers have proposed adapting the town: requalifying public spaces to be able to store the water gathered by them: storing rainwater gathered on the surface of parcels of land in urban islets; slowing, infiltrating and storing the most upstream water in abandoned areas of the hills; compensating for the existing network's shortcomings by favoring surface flows in public spaces so as to lead the water into storage areas. Technical resilience is therefore harnessed for the good of the urban project.

5.2. Applying resilience: adaptation and resistance of the material components

The resilience of urban systems involves urban programming that is adapted to risk-prone areas. However, a certain number of projects must also be based on specific approaches that focus on smaller scales, those of the material components of the urban system, as they will play an essential role before, during and after the crisis. For example, for some events that are highly diffused or affect vast areas (earthquakes, fires, terrorism, etc.), optimizing the location of buildings or of networks is not enough. For these risks, the intrinsic properties of material infrastructures must make them more or less invulnerable. In this precise context, resilience and vulnerability are tightly linked, which can appear contradictory: the more vulnerable a component, the less resilient it is.

With the rise of geomatics, some tools and methods of analysis now allow us to better identify and locate these urban components and their functions [PEU 94, EGE 98, LAR 99, THE 99]. Spatialization, simulation and 3D vision can facilitate the consideration of inherited vulnerability into the definition of urban projects.

Notably, three types of urban component can be identified which must be the focus of all efforts to create a more resilient town:

– "strategic" urban components, emergency centers, police stations, town halls, whose "functions" are to harbor the actors who deal with the emergency and provide logistical and institutional support during the crisis;

– "aggravating" urban components, such as installations classified for the protection of the environment (ICPE), hydrocarbon storage, etc. In the case of a fault, these elements lead to an amplification of the risk. It is important to know them and to act on them before the crisis, to avoid aggravating the consequences of the initial disturbance by "domino" effect (for example, pollution due to non-protected stocks, an industrial accident, etc.);

– "minimizing" urban components: refuges, for example, ensure better resilience. These components usually provide protection against risks and the disturbances that they lead to, but they can also be the cause of risks, or take on considerable damage, making the management of the emergency and the after-crisis less effective.

5.2.1. *Example of resistance to earthquakes*

In the case of earthquake risks, the new "Eurocode 8" regulations groups together six different parts that redefine the dimensions of buildings. The regulations, which govern the anti-earthquake components of buildings, aim to protect individuals, limit damage and ensure the maintenance of structures important for the protection of civilians. A map of the different levels of various hazardous events has been produced: it defines five zones and establishes different expectations depending on the event and on the type of urban component concerned. Earthquakes cover vast areas, which rules out adapting urban planning.

This piece of regulation focuses on the resilience of buildings, defined here as their capacity for physical resilience, and, as in the case of resilient planning, distinguishes between the different measures to take in function of the importance of the urban components in terms of risk management. Thus, in zones of "moderate" hazardous events, there are no obligations with regards to individual and collective houses; however, Eurocode 8 must be applied for buildings that receive the general public, such as schools or hospitals.

As long as it is followed by the adequate preparation of those involved to the risk, the application of anti-earthquake measure yields highly satisfactory results. In 2010, three earthquakes of similar magnitude (6.5 Mw) occurred in Taiwan, Bam in Iran, and in California. In California, there was no damage, but one death; in Taiwan, the earthquake caused no damage and no human loss, while in Bam, over a fault that was known but inactive for 2,000 years, and with little knowledge of the risk, 80% of the town was destroyed and close to 38,000 people died. In all three cases, the event at risk was known; however, in Taiwan and California anti-earthquake regulation is adapted and applied, while in Bam regulations exist but are not controlled, and information on the risk and training of local authorities is practically non-existent[4].

The HAZUZ 99 guide edited by the American *Federal Emergency Management Agency* (FEMA) presents detailed methods for the evaluation of risks on the urban scale. The goal is to provide an estimate of the losses that a territory could suffer during an earthquake

---

4 Winter T., *Le risque sismique en Provence Alpes Cotes d'Azur, Congrès des gestionnaires du risque sismique* – 11 juin 2009.

to provide a basis for the creation of policies regarding prevention and post-earthquake management.

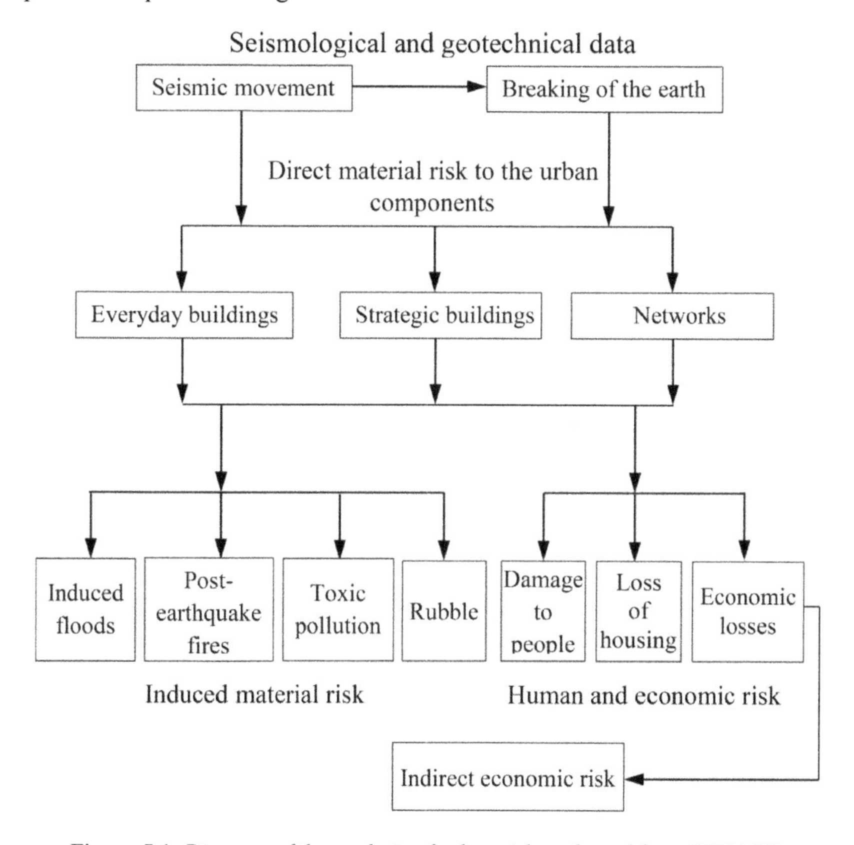

Figure 5.1. *Diagram of the analysis of urban risks, adapted from [FEM 99]*

Beyond the estimation of risk, the capacity to maintain oneself over time despite the occurrence of potentially damaging events, just like the capacity to cope with these events, involves the designing of resistant buildings that limit material, human and economic damage in the case of a major event. The study of the resistance of urban components traditionally combines not only the design characteristics of the structure, both in drawing and in construction, but also the quality of the materials, the age of the structure, its height, etc. For example, feedback shows that for the same outline, the structure of a building is more vulnerable with dry stone masonry than with concrete, and steel has even better results.

In the case of new constructions, the behavior of important structures (bridges, roads, industrial installations, gas distribution networks, electric networks, etc.) in the face of an earthquake requires previous structural calculations, using mathematical models and experimental methods. However, without raising awareness of the risks in the actors involved, and particularly in those involved in urbanism and in urban planning, these measures are not applied.

Awareness must also be raised in users and in the general public. Beyond simple knowledge of the right things to do during a crisis, for earthquakes this involves adapting households and taking precautions with regards to the positioning of objects that are likely to fall, especially heavy objects and those that can break, such as lamps, mirrors, bottles, etc. Special care must be taken with toxic or inflammable products so as to avoid spills in the case of a fall [ISA 12].

The environment must also be taken into consideration in the construction of new buildings. Feedback from previous experience often shows constructions that did not resist the earthquakes due to the instability of neighboring buildings. It is still difficult to plan for the protection of new urban components while taking into consideration the fragility of neighboring constructions, but this study of proximal interdependences is vital, nonetheless.

It is also crucial to increase the resistance of existing urban material components. Upgrading buildings to make them earthquake-resistant requires first an estimate of the vertical loads (mass of the structure, of the exploitation and of temporary events like snow) and then the horizontal loads (effect of wind and earthquake) of each part of the building. The importance of the movement of the ground during earthquakes is not necessarily a determining factor as the distribution of masses and of rigid elements of the constructions can amplify or dampen the oscillations imposed on it at the level of the ground.

The measures taken to be able to reach the desired level of performance exist in several orders; for example, it is possible to carry out the earthquake-proof isolation of a building, or to block its resonance. This solution is usually very costly, and only done for urban components that are strategic during a crisis. It is achieved by a remodeling of the ground-building transition. To stop resonance, which amplifies the effects of seismic waves, several routes are available: reducing the various weights by replacing floors, covers, etc., or using lighter elements, is one part of the response. Increasing the rigidity of the construction is also an effective strategy to stop resonance, but it is effectively a reinforcement of the structure. Finally, a relatively recent technique involves the use of anti-earthquake dampeners, usually placed in the bracing elements.

A building can also be provided with enough mechanical resistance by greatly reinforcing the different elements of the main structure. However, this primary mode of action for new buildings is harder to put into practice for existing ones.

Finally, it is possible to transform the component with the goal of encouraging its dissipative properties, i.e. its ability to cope with deformation without collapsing. This solution is advantageous as it allows a reduction in the level of mechanical resistance of the structure. However, for an existing building, it is not often possible to provide it with a level of structural resistance that is equal to that of a new building.

The cost and technical feasibility of the measures are important decision criteria, as the strategy that is adopted is usually a compromise between the cost of the upgrade, and the amount of accepted damage. Techniques also exist for modifying the grounds, leading to risk reduction.

5.2.2. *Example of resistance to flooding*

Proposals for making a town more resilient to flooding are usually based on a two-sided strategy that involves retracing the water pathways on the scale of the shape of the town, and limiting impacts on urban components on the scale of the building.

Resilience is first thought of on the scale of a district or of the outline of the town. Streets, parking lots and various intermediate spaces can, depending on the existing and/or corrected topographies, enable the temporary circulation of water. The Hoche Sernam district in Nîmes was remodeled in this way, starting in the 19th Century with the installation of equipment that could not be built in the existing town: a hospital, workshops and warehouses for the SNCF, and a fire station. Floods caused by runoffs are significant in this area that suffers from *cévenoles* rains and is located in the topographical context of a hillside. Thus, in 1988 (reference event of *supra* centennial occurrence), water heights of more than 1 m were recorded. Nowadays, from an urban strategy perspective, Nîmes is looking to densify itself. The Hoche Sernam district, while floodable, is a district that benefits from significant land reserves: it could be the recipient of many programs allowing the town to develop itself from within.

In the winning answer, the level of roads is lowered to carry water to a large urban park. The park is organized around ponds to slow flows. Light structural devices (small walls, embankments) protect the neighboring Richelieu district, which is currently highly exposed. The site of the project covers an area of 20 ha, but the model of the water flow (squares grid with 0.5 m sides) is done on the scale of the study perimeter, which is 150 ha. In the case of an exceptional freshet, such as in 1988, the maximum height of the water at the base of the construction had to be lower than 20 cm – doorways were therefore located above this point. Currently, architects are working on the adaptation of an area of 130,000 m^2, to comply with programs of several thousand homes and a university. From the perspective of the phasing of the execution of the project, it is necessary that the builders respect a master plan that is tightly linked to the functioning of the flow axes.

Romorantin-Lanthenay is another example, located in the department of Loir-et-Cher, and is a commune located on the banks of the river Sauldre. A project is currently in operation on one of the banks, and includes individual houses, collective housing, local facilities, as well as parks. A singularity of this layout is that the streets, public spaces and buildings facilitate the river water to pass freely. The individual houses are built on stilts, reducing risk and aiding the passage of water. The buildings are oriented so as not to form a barrier against the river water. On top of these various arrangements, a reservoir basin has been placed in the main park.

Resilience is next thought of on the scale of buildings. Water can penetrate in numerous ways: by climbing up from the water table, travelling through walls: through openings: through networks (connective conduits or more generally the sewage network). To deal with this, resilience can be based either on the "dry" method, or on the "humid" method.

The "dry" method, also called "dry flood-proofing", is a waterproof method that maintains the water-tightness of the building by stopping water from getting in. The principle involves isolating the outside as much as possible. The techniques use the peripheral walls of the construction as "dams". The water-tightness of the walls is required, as well as the presence of devices that allow openings to be closed. This "dry" method is particularly adapted to floods of less than a meter, and with submersion times that are relatively short.

Another technique consists of an elevation above the flood level, either permanently – by lifting the construction onto a backfill, crawlspace or on stilts – or temporarily, as in the case of floating constructions. The floating constructions can be carried out using current floatation devices. A French society is developing a floatation device called Batifl'O, made of Impermeable High Density Polyethylene, with a hollow center into which a concrete slab is fitted. Batifl'O can measure up to 2.44 m × 2.44 m, with a height of 1.1 m. The floatation system with its concrete slab can support 1.5 tons/m^2 and operational charges of 250 kg/m^2, which covers all current buildings. Tests show that the buildings resist heights of 2.8 m and speeds of 2.78 m/s. There is no limit in terms of surface area, as the floaters can be linked together and create platforms of several hundred square meters. These slide vertically on pylons that ensure a connection to the grid. With such properties, the floaters can support individual constructions or small collectives of two or three stories.

The floaters also enable access for all vehicles, as the path is designed identically to a path on a backfill, except the backfill is replaced by the floaters. For existing constructions located in floodable areas, floaters can provide platforms within these for the storage of high value material, polluting products, etc.

The "humid" method, or "wet flood-proofing", however, lets water enter the building, without damaging the structure of the construction that was designed for this occurrence, and without touching goods located inside, placed out of the reach of water (by modifying them or by using floating platforms). The materials used in construction must be able to dry out quickly and conserve their properties after the flood. This method presents advantages for high water levels, where complete water-tightness can weaken the stability of the building through pressure of water on the walls and buoyancy-induced pressure on the building as a whole.

Thus, in Venice, the water of the *aqua alta* runs into the Piazza San Marco and inside many of the palaces. The Querini Stampalia foundation, restored by Carlo Scarpa between 1961 and 1963, lets water flow into the building. The architecture ensures the rooms may still be used by raising certain parts of the floor, which are used for moving around in the case of flooding. The choice of adapted materials minimizes inconvenience caused [LOS 98]. Water is no longer a problem, but rather a theme and a source of inspiration: it runs freely. The floodable nature of the site magnifies the project, creating central architectural references.

5.3. Conclusion

The resilience of the town, understood here as adaptations that aim to develop the capacity for resistance of material elements, must therefore rely on a multi-scalar approach: it is seen at the macro scale of the town with urban planning, at the meso scale of the project, and at the micro scale of the urban component and of what it contains.

The resilience strategy completes those of reduction and of anticipation of the hazardous event. It involves an active appropriation of the risk by the relevant actors, in particular by populations, and the development of preventative actions (monitoring, alarms, etc.). Therefore, works of urban planning cannot be disconnected from the organizational level mentioned in the previous chapters.

Current changes in urbanism and urban planning will certainly change practices. The recent appearance of "project urbanism" will help in the realization of reorganization projects and in the coordination of actions and linking regulatory procedures as part of complex projects involving urban territories considered at risk. These changes of urban planning and organization also provide the means to put into practice approaches that integrate risk into urban projects. While experiments have been limited in size and only exploratory in nature, they have already shown their relevance, far beyond their current applications. The balance between local actions, standards and national procedures will undoubtedly need to be redefined.

5.4. Bibliography

[BEU 08] BEUCHER S., MESCHINET DE RICHEMOND N., REGHEZZA M., "Les territoires du risque. Exemple du risque inondation", Historiens et géographes, numéro spécial *Construire les territoires*, pour le congrès 2008 de l'UGI à Tunis, no. 403, pp. 103–111, 2008.

[EGE 98] EGENHOFER J.M., GOLLEDGE G.R., *Spatial and Temporal Reasoning in Geographic Information*, Oxford University Press, Oxford, 1998.

[EME 07] EMELIANOFF C., "La problématique des inégalités environnementales, un nouveau paysage conceptuel", *Écologie et Politique*, no. 35, pp. 19–31, December 2007.

[FEM 99] FEMA, *Earthquake Loss Estimation Methodology HAZUZ 99*, Federal Emergency Management Agency, Washington D.C., United States, 1999.

[HER 09] HERNANDEZ J., "The Long Way Home: une catastrophe qui se prolonge à La Nouvelle–Orléans, trois ans après le passage de l'ouragan Katrina.", *L'Espace géographique* vol. 38, no. 2, pp. 124–138, 2009.

[ISA 12] ISARD, Information Sismique Automatique Régionale de Dommages, available at http://isard.brgm.fr, accessed, March 2012.

[LAR 99] LARDON S., LIBOUREL T., CHEYLAN J.-P., "Concevoir la dynamique des entités spatio-temporelles, Représentation de l'espace et du temps dans les SIG", *Revue internationale de géomatique*, vol. 9, pp. 45–65, 1999.

[LOS 98] LOS S., FRAHM K., *Carlo Scarpa*, 1998.

[NOV 02] NOVEMBER V., *Les territoires du risque: le risque comme objet géographique*, Peter Lang, 2002.

[NOV 11] NOVEMBER V., PENELAS M., VIOT P. (eds), *Habiter les territoires à risques*, PPUR, Lausanne, 2011.

[PEU 94] PEUQUET D.J., "It's about time; a conceptual framework for the representation of temporal dynamics in geographic information systems", *Annals of the Association of the American Geographers*, no. 3, pp. 441–461, 1994.

[THE 99] THÉRIAULT M., CLARAMUNT C., "La représentation du temps et des processus dans les SIG: une nécessité pour la recherche interdisciplinaire", Représentation de l'espace et du temps dans les SIG, *Revue internationale de géomatique*, vol. 9, pp. 67–99, 1999.

[WIN 99] WINTER T., Le risque sismique en Provence Alpes Cotes d'Azur, Congrès des gestionnaires du risque sismique, 1 June 2009.

6

Adapting Territorial Systems Through their Components: The Case of Critical Networks

The question of resilience as an adaptation to socioterritorial systems has found its place in the drive to make cities more resilient. This injunction comes from a slow rise in awareness of the vulnerability of urban spaces. As a result, the issue of resilience is particularly important when applied to towns.

Urban spaces, and large cities in particular, are now seen to be highly vulnerable places. Urbanization has led to an increase in the concentration of men and goods, increasing the potential for damage in times of crisis. Urban sprawling leads to the occupation of high-risk areas, previously left unoccupied, whereas urban technical networks are no longer properly adapted and prove to be too small, leading to new sources of danger and new forms of damage. Finally, urban spaces have become increasingly complex sociotechnical systems, when the overlapping of activities, networks and territories has led to the very rapid diffusion of disturbances on very big scales.

We can also note the appearance of new threats: either new events (electric power cuts, internet networks failures and nuclear power outages), or old events (floods, storms, industrial hazards and

---

Chapter written by Damien SERRE.

terrorism) whose dynamics have become complex as a result of the domino effect. We can also look at the impact of climate change on the frequency and intensity of certain events (cyclones, rise of sea levels, etc.).

Therefore, it becomes necessary to develop new managerial strategies to anticipate scenarios that probabilistic models have judged to be extreme or rare and to cope with the inherent uncertainty of this new context. The goal is to adapt urban systems to this new state of affairs by considering several spatial and temporal scales. The resilience of an existing town in the face of risk must be increased, but new urban districts must also be built, or even new cities. Some urbanists even advocate the transformation of hazardous events, seen as negative occurrences, into urban opportunities.

6.1. Technical and critical networks, strategic elements of resilience

The building of a resilient town relies on several levers. One of them in particular has caught the eye of urbanists and planners: the adaptation of technical networks.

These networks are indeed essential components of the urban system. Their importance has increased over time, as they link an increasing number of people and places, creating in this way situations of interdependence that, during normal periods, provide various resources and opportunities, but in times of crisis can represent major factors of vulnerability. Feedback analysis shows both a dependence of the urban system on its technical networks, and a diffusion of the effects of events via these same networks. Various studies on urban technical networks show that these are both vulnerable to hazards and propagators of disturbances due to their interdependences and their extensions [VIG 06]. They also play a particular role in the occurrence and development of disasters, participating equally in the damage process and in the management of the crisis and the recovery process [FEL 05].

6.1.1. *Technical networks in towns*

The development and the dependence of towns on urban technical networks results in a certain type of urbanization, called reticular urbanization. The organization of networks is not only limited to that of a technical system, but also interferes with the spatial organization of the city. This type of urbanization is at the origin of the diffusion of risk in urban spaces.

Indeed, the development of technical networks starts to intensify in the middle of the 14th Century. Networks quickly become increasingly complex development axes that form interdependent meshed structures. Throughout the 20th Century, public transport, electricity and gas distribution networks, telephone, fiber optic, central heating and waste management networks and others have made the structure of this mesh more complex, as well as complicating interdependence relations.

Nowadays, urban technical networks are very vulnerable: they have a high potential for damage. They are also a source of vulnerability on the scale of the urban system, as the functioning of a town depends largely on their proper functioning. Even a minor fault can have downstream effects on all of the system.

The effects hurricane Sandy had on the city of New York in November 2013 clearly illustrate the sensitivity of networks and the cascade of disturbances produced by their fault. Sandy caused the destruction of part of the electrical network: overhead contact lines destroyed by the storm, but also, in Manhattan, the infiltration of seawater into the underground network. In New York, 250,000 homes lost electrical power for several days (up to a week, seeing as each building then had to be checked one by one). Daily life was greatly disrupted on the scale of the whole city, with the closure of the underground due to a lack of electricity and partial flooding, but also the closure of some buildings, notably very tall ones (elevators, heating, security systems all out of order). Telecommunications networks (cellular and internet) were also greatly disrupted. The closure or slowing of refining plants led to a lack of fuel, needed for the circulation of emergency vehicles and emergency generators. In the flooded neighborhood of Breezy Point, electric cables that had come into contact with salt water caused a fire that quickly grew in size due to the proximity of the houses to each other and the fact that the fire trucks were either damaged or blocked by the flooding of the roads.

The failure of technical networks shows that they are sensitive to all type of hazardous events, from human error to terrorist attacks, through technological and natural events. Managers agree that urban technical networks must be made safer in the 21st Century to avoid this type of failure. However, due to the overlapping of networks, their tentacle-like extension and the extreme density of certain knots, increasing the resilience of networks is very difficult.

6.1.2. *Technical networks: critical infrastructures*

The analysis of the behavior of urban technical networks leads to the identification of two important notions that are linked:

– the notion of "critical infrastructure". "Critical" infrastructure can be defined as a set of installations and services necessary for the function of the city [ASC 09]: its failure threatens security, the economy, the way of life and the public health of a town, a region, or even a State;

– the notion of the interdependence of networks: most critical infrastructures interact with each other. These interactions are often complex and unknown, as they go beyond the limits of the system in question.

As part of the analysis of interdependent critical infrastructure, a distinction can be made between two types of interaction:

– interactions with one single critical infrastructure (the energy network, sewage network or road network);

– interactions between critical infrastructures [MCN 07], which requires an analysis of a network of networks (the macro-network).

Therefore, analysis of interdependences requires a change in scale to analyze the components of a system (fine scale), followed by the relations between systems (bigger scale of the meta-system): critical infrastructure is first analyzed as a system in of itself, and then on a more encompassing scale, such as a system of critical infrastructures (macro-network).

As an example, in the case of a flood, interdependent critical infrastructure can lead to the following scenario:

– heavy rain over a densely populated and urbanized area causes flooding through an overload of the rainwater networks;

– the transport networks are then disrupted by the flooding of their roadways;

– this immersion causes disruptions in the movements of people and in general, threatening economic activity;

– communication pathways, acting like a river bed, not only facilitate the water to reach houses but also the entire infrastructure that ensures essential urban functions (administrative buildings, shops, health centers, etc.);

– energy and telecommunications networks can in turn be affected by the flood and be led to dysfunction, which paralyzes daily economic life even more.

Here again, the example of Sandy is enlightening: the sewers overflowed and around 40 billion liters of dirty water spilled out onto the streets of New York and New Jersey, leading to a halt of road traffic, while at the same time the subway system was flooded with seawater. These untreated waters are a considerable risk in terms of sanitation and the environment. In Maryland, the overflowing of a water treatment plant led to a power outage. In addition to dirty water, household waste spilled out onto the streets.

6.2. Choosing adaptations

Technical networks have therefore been identified as entry points for faults, and are now systems on which technical and managerial measures are likely to focus. Evaluating the resilience of urban technical networks appears to be an essential phase for increasing the resilience of cities and for orienting responses to be put in place to decrease the effects of hazardous events, whether this is through improving networks and evacuation plans, or better defining prioritized interventions. The capacity of these networks to work in a damaged state after suffering the disturbance must be evaluated, as well as their capacity to be returned to service, to allow the city to recover as quickly as possible. This evaluation notably leads to the identification of possible dysfunctions of the technical networks and the diffusion of deleterious effects, following the occurrence of the hazardous event.

6.2.1. *Modeling the urban system*

Urban technical networks propagate disturbances throughout the city through a cascade of dysfunction. To evaluate urban resilience in the face of these risks, it is vital to take into account these interactions between networks, which requires an understanding of how the urban functions.

A city can be considered to be a system composed of several subsystems. The literature – which his sizeable – confirms the importance of systemic analysis to study urban areas. These works have already given rise to complete models, which, however, are hardly applicable to the issue of risk. This is why a new model is being proposed.

To build this model, we must identify the components of the urban system, which, if damaged or disrupted, would most affect the operation of the town. This goes from the assumption that the evolution of urban systems is tightly linked to economic activity, which highlights the role of businesses. Nowadays, the importance of economic activity is outlined by the attraction that a town has on the populations of another town, particularly due to competition between job markets. Housing and equipment are also essential urban components for receiving populations and for economic activity.

However, it is important to separate these components of the urban system into two separate categories:

– technical components – in this case the networks – which can themselves be seen as systems;

– infrastructure that enables the proper functioning of urban society, hosting decision-making organs (town hall), "sovereign" functions (police, justice) and non-profit services (education and health).

These two categories are opposed by nature: linear/punctual, technical/social, contained/containing. However, their involvement and interrelations are essential.

The physical environment must also be included in our model. The urban system is developed in constant interaction with it: the physical medium therefore can be neither separated from the urban system nor can it be reduced to a simply supportive element, or considered as external to the town.

This results in the following model (Figure 6.1).

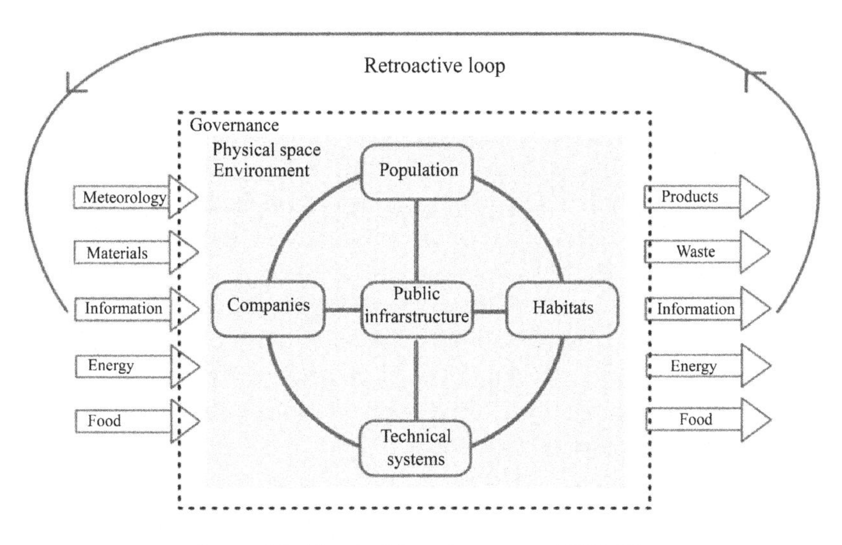

Figure 6.1. *Model of the urban system [SER 11]*

Moreover, the city is considered to be an open system, which requires us to look at the relations that the city has with surrounding spaces. For example, the relations that the city has with its hinterland are very important as they allow it ensure the steady arrival of supplies. The city receives a certain amount of primary matter and foodstuffs for its own use, but also needs to ensure proper conditioning and production of finished products, to be traded with other cities. Exchanges with other cities are the second most important type of relation with the system's environment.

One of the main productions of a city, and a major issue, is waste. This type of production highlights a characteristic of systemic analysis: retroaction. Indeed, waste can be considered to be elements exiting the system, but this does not mean that they are without consequence on the city and its biophysical environment. Another element that is essential when dealing with risk is the "meteorology" box, insofar as many hazardous events are more or less directly linked to rainfall and temperature.

The influence of higher-ranked decision-making organs makes the model more complex. It is therefore important to represent supra-urban levels of decision-making, the term chosen here being governance.

Finally, we must deal with the issue of the spatial limits of the system. These limits are hard to establish as they vary depending on the question at hand. The urban dynamics mentioned previously have mostly blurred the traditional boundaries between that which is urban and that which is not, between different types of territoriality (economic, societal and political), creating multiple entities whose limits are not the same. In a quest for clarity, and seeing as management operates on the level of administrative parameters, we have chosen communal limits, although we acknowledge the arbitrary and incomplete nature of this choice.

6.2.2. *The role of networks in the urban system*

By considering the city as a system, the interrelations between the different urban components are highlighted. As a result, it is interesting to study these interrelations following a disturbance, starting with the urban model created. It is possible in this way to create a model of the urban system during a crisis.

It would appear that, due to the limitations of installation and of their very structure, networks are not only the “entry point” of the hazardous event (initial source of danger), but also the “entry point” of the risk (all of the disturbances and damage) considering the domino effects that can be induced by these networks.

The risk is propagated through these networks following various possible scenarios. We can then propose a systemic approach for the behavior of the city in the face of risks.

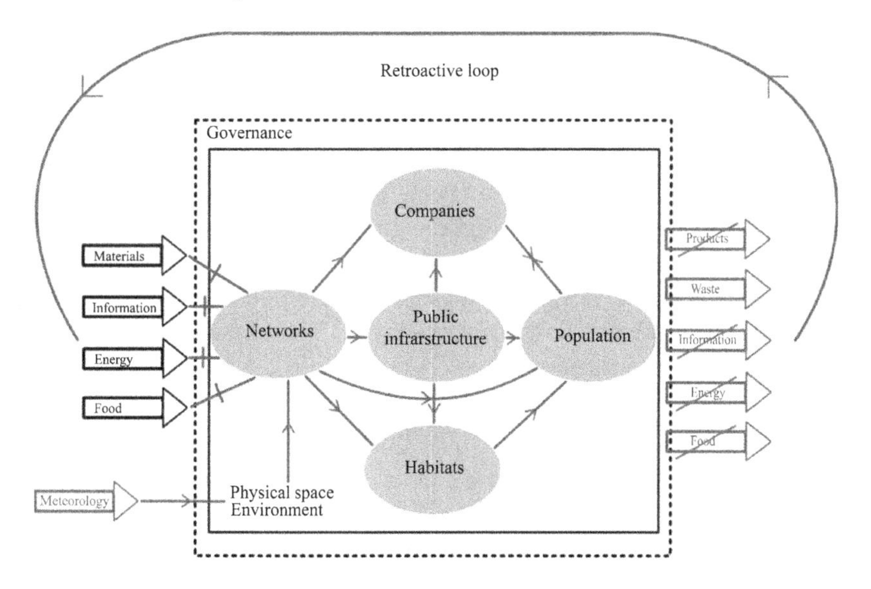

Figure 6.2. *Technical networks and the propagation of flood risks in town [SER 11]*

To build a resilient town, resilient buildings are not enough: above all, technical networks must be available that are able to withstand all impacts, absorb them and recover. These three properties that define technical resilience are vital:

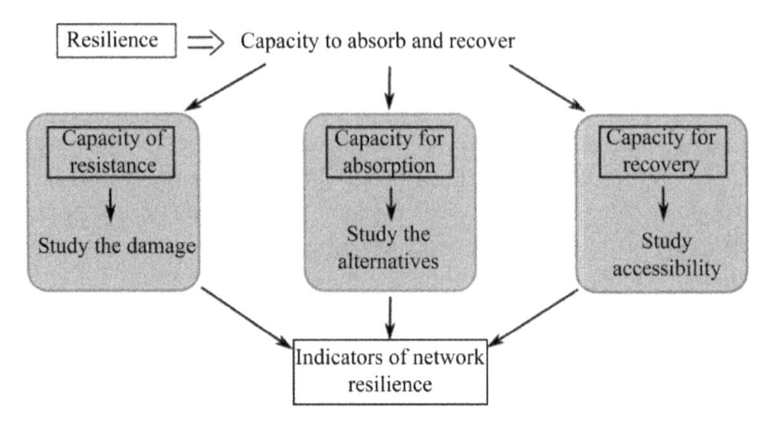

Figure 6.3. *Capacities to be studied for networks resilience [SER 11]*

1) The capacity for resistance when facing a disturbance

The capacity for resistance goes back to the material damage of the network following a hazardous event. The more a technical system is materially damaged, the higher the chances of the system malfunctioning, and the harder it will be to return it to normal operation. For this, we use methods taken from operational safety (see infra) to determine the damage taken by the system and to take interdependences into account.

2) The capacity of absorption when facing a disturbance

The capacity of absorption depends on the alternatives offered by the network following a failure of one or several of its components. For example, when a transport network is damaged, the traffic splits off onto alternative paths from the initial one. The greater number and higher variety of paths there are, the less the disturbance is felt [GLE 07]. These alternatives facilitate the service to continue and allow the network to continue operating when damaged. Here we must study the configuration of the network so as to characterize its redundancies. Methods taken from graph theory provide interesting answers.

3) The capacity of recovery

Recovery is an essential capacity for the resilience of system. For a network, this recovery can simply be the time needed for a return to service of one of the damaged components. Here, purely technical aspects are linked to more organizational aspects. Recovery is also involved in the accessibility of services that allow the network to return to normal and of the components that might have been damaged. The goal is to use the elements of spatial analysis rather than the organizational elements that require a lot of information: geographical information sciences help in providing evaluations of the capacity of recovery.

6.2.3. *The tools at our disposal*

To carry out the diagnosis of a work of civil engineering, such as a network, evaluate its state and performance, or even to analyze the risk and evaluate its level of operational safety, two types of approach are possible [ZWI 95]: internal methods and external methods. "Internal" methods rely on a deep knowledge of the system studied (for example, a set of interconnected industrial components, or a work of civil engineering). From models, it is possible to analyze the mechanisms of damage and of rupture, and to predict future behavior.

Depending on the type of model used to describe the system, a distinction can be made between two types of modeling:

– physical modeling relies on the physical representation of continuous or discrete processes of damage of the system, taking into account the equations that govern internal phenomena. It requires considerable knowledge of the system, as well as its representation in the form of physical and mathematical models, supported by digital simulations;

– functional modeling: the systems are studied from the angle of the functions that they must carry out and for which they were designed. The principle of functional modeling involves determining what the interactions are between the components of a system and its environment, so as to formally establish the links between the failures of functions, their causes and their effects.

"External" methods can be applied to contexts where the modeling of mechanisms (physical or functional) is not technically possible, or is not adapted considering complexity or cost. Depending on the information available, we can distinguish between the methods based on statistical analysis and those based on expertise [PEY 09].

Moreover, urban technical networks can be defined as complex systems. By this is meant a system composed of a large number of elements that interact between themselves in a nonlinear fashion [SIM 91] and in which the relations of cause and effect are not always established scientifically and require a certain amount of intuition. Methods of operational safety have been developed to study complex systems (high numbers of components, multiple and looped faults, etc.) for which it was very difficult, or even impossible, to produce an operational model using classical physical approaches. First used in industrial systems, they are now applied to technical urban networks.

Several methods are applied:

– functional analysis: the system studied is modeled and we are looking for the functions that it carries out;

– the analysis of modes of failure and their effects: we are studying here the effects of the failure of each component of the system, considered successively.

Unlike classical studies of risk analysis in civil engineering, where usually only one type of work is used, this approach is original in that it is used over all urban technical networks so as to take into account their interactions [SER 05].

6.2.4. *The development of decision-making tools*

Once the evaluations have been carried out, decision-making tools must be made for networks managers, and more globally for all of the urban actors more or less involved in the management of risk.

There is a double objective here. The first objective is to have tools that can be connected to indicators that help make decisions. The possible joining of geographic information systems (GISs) with indicators of resilience must be looked into. The second objective is for these tools to be able to be shared by various actors interested in the improvement of urban resilience: the help of client/server technology via the internet is an interesting option in this case.

The use of GIS in risk management has become widespread in all those involved for two main reasons. First, GIS have become popularized, as they are now technologically and economically accessible. Second, the spatial dimension of risk is increasingly recognized, and GIS lead to a form of risk spatialization.

In parallel with these developments, the theoretical framework of risk management has been continuously evolving, just like GIS tools. The main vocation of GIS, which aims to "regroup, with a single tool, various but localized data in the same geographical space, relative both to the Earth and to Man, to their interactions and their respective changes" [DEN 04], remains applicable. However, the development of 3D, web and real-time applications, or even more generally the development of computer tools and databases, lead to the improvement of the "gathering, management, analysis, modeling and presentation of data of spatial reference so as to resolve complex problems of planning and management" [DEN 04].

6.3. Conclusion

Adapting an urban system to make it more resilient involves considering all of the components of the urban system and their interactions. On top of the systemic approach used to deal with urban functions in a time of crisis (reorganizing the town and its technical networks), which already combines engineering (technical solutions and technology) and organizational points of view (political choices and management modes), there is also the point of view of social sciences which focuses on governance, populations and their significance, those involved in towns and in services.

The focus on resilience must not be seen as a limitation imposed from above or from the outside. However, it must be developed in collaboration with local collectivities and local managers, who know the territory, its issues, its potential and its limitations. This is why the tools developed to evaluate the resilience of networks must be accessible to non-experts, or at least allow for the creation of models that support dialog and discussion between different services and managers.

However, this analysis of the resilience of networks requires a large investment from networks managers, notably with regards to the data needed to create the model. It is therefore vital to make those involved aware of the advantages of such a collaboration. Here, we first of all propose a mutual form of learning of the existing links of interdependence within the territory between the systems. A simple self-diagnosis by the manager must first identify the resources that are useful for the functioning of the service and put forward the dependence of the system on other systems. In the same manner, the identification of users dependent on resources produced by the system must make the manager aware of the importance of the continual operation of their service. The pooling of sectorial diagnoses highlights the complexity of the interactions between systems as well as possible distortions of perceptions: through the characterization of the critical nature of the resources, we can pick up neglected or under-estimated links.

This first step of confronting interdependences in the presence of all those involved in the towns and networks precedes a more developed analysis, which, however, must always operate in collaboration with the actors on the ground, who must adopt its approach. This integrated approach of urban resilience must therefore be considered in the everyday management of a territory and originate in local authorities that know its ins and outs. At this point, it is no longer seen as a constraint, limiting possible urban planning or forcing sizeable and costly works to be carried out. The costs and benefits of each measure is evaluated alongside the knowledge of impacts that they would have on the global system, on the populations and activities serviced. All of those involved are joined around an analysis that seeks to favor a long-term vision.

6.4. Bibliography

[ASC 09] ASCE, Guiding principles for the nation's critical infrastructure, Technical Report, 2009.

[DEN 04] DENEGRE J., SALGE F., *Les systèmes d'information géographique*, Presses Universitaires de France, 2004.

[FEL 05] FELTS L., Vulnérabilité des réseaux urbains et gestion de crise, CERTU Report, 2005.

[GLE 07] GLEYZE J.-F., REGHEZZA M., "La vulnérabilité structurelle comme outil de compréhension des mécanismes d'endommagement", *Géocarrefour*, vol. 82, nos. 1–2, available at http://geocarrefour.revues.org/1411, 2007.

[MCN 07] MCNALLY R.K., LEE S.W., "Learning the critical infrastructure interdependencies through an ontology-based information system", *Environment and Planning B: Planning and Design*, vol. 34, pp. 1103–1124, 2007.

[SER 05] SERRE D., Evaluation de la performance des digues de protection contre les inondations, PhD Thesis, Marne-la-Vallée University, 2005.

[PEY 09] PEYRAS L., Evaluation de la performance et des risques des ouvrages hydrauliques de génie civil, PhD Thesis, University Blaise Pascal, 2009.

[SER 11] SERRE D., Flood resilient city – assessment methods and tools, PhD Thesis, University Paris-Est, 2011.

[SIM 91] SIMON H.A., *Sciences des systèmes, sciences de l'artificiel*, Dunod, 1991.

[VIG 06] VIGNERON S., DEGARDIN F., Réduire la vulnérabilité des réseaux urbains aux inondations, Internal report, CERTU, 2006.

[ZWI 95] ZWINGELSTEIN G., *Diagnostic des défaillances. Théorie et pratique pour les systèmes industriels*, Hermès, 1995.

7

Resilience and Global Climate Change

Until the past few years, scientific research on climate change has largely ignored the notion of resilience, just as it has paid only scant attention to the city as an analytical scale of the "impacts" of climate change and as a grounds for applying policies for limiting greenhouse gas (GHG) emissions. It is sufficiently convincing to consult the last Intergovernmental Panel on Climate Change (IPCC) report, dating from 2007 but essentially prepared between 2000 and 2005 [IPC 07]. The section of the report focusing on the WG2 *(Impacts, Adaptation and Vulnerability)* only mentions the concept of resilience in relation to the resilience of ecosystems or indigenous peoples in the Arctic. The impacts of climate change on the city do not form the object of any particular study and only appear in passing in case studies focusing on Hurricane Katrina and on the 2003 heatwave in Europe and France. In the section of the report focusing on WG3 *(Mitigation)*, policies for reducing GHG emissions are analyzed by sector (energy, transport, housing, etc.), in a macroeconomic approach on a global or national scale, and without any explicit reference to the city.

Similarly, the concept of urban resilience, arising from disciplines that focus on analyzing natural, technological or social risks, seems for a long time to have remained outside the prospective, if not predictive, dimension that characterizes research dedicated to global

---

Chapter written by Claude KERGOMARD.

change. The retrospective or geohistoric approach and the importance accorded to the experience feedback in the analysis of factors for urban resilience seemed at odds with the handling of uncertainty as practiced within the framework of predictive models and scenarios developed for global change. Urban resilience should however provide food for thought on adaptation to climate change, or the transition of energy supplies imposed both by the necessity of creating a "carbon-free" economy and by the scheduled depletion of fossil resources.

For some years however, the situation has been evolving, partly on the initiative of those cities and their institutions, who themselves signed the Global Cities Covenant on Climate (Mexico, November 2010). In an extension of the work of the IPCC, an international network of researchers on climate change and cities (*Urban Climate Change Research Network*) came into being in 2008; its first report [ROS 11], without explicitly adopting the term "resilience", emphasized the combined analysis of risks caused by climate change, urban vulnerabilities and capacity for adaptation.

Similarly, the increasing influence of a literature that uses the term "resilience" to mean urban adaptation to global changes [LEI 11, NEW 09] is being promoted. This literature often uses resilience as a substitute for, or a furthering of already dated approaches to, lasting or "sustainable" urban development. Since it wishes to convince a broad public audience without straying too far into theoretical discussions, this corpus questions only small implications and limits of this reconciliation; it often displays the weakness of concentrating on only one aspect of climate change – the increased probability of "extreme" climate events. Therefore, it is necessary to pursue further the potential role of the concept and practices associated with "urban resilience" in adaptation to climate change and the transition of energy supplies, by successively considering:

– the issues of scales, temporalities and the handling of uncertainty posed by reconciling global changes and urban resilience;

– the concordances and discordances between the concept of urban resilience on the one hand and, on the other, the way in which scientific research on climate change treats adaptation to its inevitable effects and to the new energy regime imposed both by depleting fossil resources and reducing CO_2 emissions;

– the practice and application of "sustainable" town planning and its compatibility with urban resilience in the sense that this work wishes to promote. Are eco-districts or "sustainable" cities as they are promoted today also resilient cities?

7.1. Resilience and global change: scales, temporalities and uncertainty

The problem of climate change, and global changes more generally, emerges from the combination of the concerns of the scientific community of climate physicists and those of decision makers in international organizations who, centered around the UN, aim above all to reduce global inequalities in development on a global scale. This combination materialized during the "Earth Summit" in Rio (1992) and led to the United Nations Convention on Climate Change, which today still establishes international policies in this domain.

7.1.1. *Global change and the city: a problem of scale?*

In this context, the city does not appear *a priori* as a priority theme: in the framework of climate system modeling, urban areas, which only represent around 1% of land surfaces, are almost negligible. Emphasis is placed on processes likely to be associated with climate change though simply causal links and which can form the subject of modeling with a physical basis: disturbances in ecosystems and natural resources, agricultural production, the rise in sea level, public health issues on a global scale, etc.

The reintroduction of the urban scale in the problem of global climate change results from two observations, which have only really made themselves felt since the second half of the 2000s: widespread urbanization (since 2007, the world's urban population has exceeded its rural population), and the emergence of 50 very large urban regions (with more than 5 million inhabitants).

This finding requires us first to take account of two relatively neglected aspects:

– the proportion of GHG likely to be attributable to urban areas greatly exceeds the only section of the world's population that they concern. The concentration of industrial and tertiary activities, the energy needs brought about by transport, lifestyles and the working of the "urban metabolism", requires urban areas to be included in policies for limiting GHG emissions, just as in *scenarios* for the "transition of energy supplies" associated with the probable depletion of fossil resources;

– the approach to the consequences of climate change using vulnerability, which is gradually making itself felt beside the mechanistic notion of impact, underlines the prominent position of urban spaces among the areas subject to climatic risk. Urbanization is itself a factor in urban vulnerability, which climatic risks share with other natural risks likely to affect large towns. "Landmark" events, such as the major 2003 heatwave in Europe or the consequences of the tropical cyclone Katrina in New Orleans in 2005, have contributed greatly to prioritizing climatic risks connected to "extreme" phenomena, but whose link with climate change caused by human activities is not easy to prove scientifically.

This evolution highlights the limits of an exclusively "globalist" (or planetary) approach to climate change, or the "*downscaling*" approach (which moves from the planet's climate to local impacts), approaches which nevertheless privilege climate system modeling. The urban organism, the site of complex interactions between social habits, energy uses and consumption methods and production of GHG, all vulnerabilities in the face of climate events still too often perceived as exogenous, becomes a pertinent analytical scale, open to policies for adaption and for increasing resilience.

7.1.2. *Global change, disasters and resilience: questions of temporalities*

The dominant representation of anthropogenic climate change, associated with the concentration of GHG, is provided by the evolutionary graph of the world's temperature. Of man-made size, its growth can be reconstructed over a century and a half of comparable meteorological observations (1860–2010), the global temperature is also the data most easily simulated for the century to come with the help of climate models, depending on the evolution scenarios for GHG emissions [BRA 09].

This method of representation in the form of a continuous magnitude, whose evolution can only be measured or modeled in stages over the course of a decade if not a century, relegates to second place the forms of climate variability over shorter timespans. The links, too often mentioned in the media, between the rise in global temperature and the most spectacular effects of climate variability (the 2003 heatwave in France and Europe, Atlantic storms such as those of December 1999 or Xynthia in 2010, sustained rainfall which has led on several occasions to spectacular floods in Central Europe), incites caution among climatologists since the link between an event of this type and the reality of global warming cannot be proven by the models currently available to the field.

Faced with the difficulty of establishing a link between the climate "disasters" observed recently and current climate change, the consensus of an overwhelming majority of the scientific community, as expressed by the IPCC reports, is that planetary global warming is of a type to lead to an increase in climate variability in a short space of time including the frequency and intensity of "extreme events" likely to cause climate disasters with, however, a hierarchy of event types: the increase in the frequency of warm heatwave-type episodes (counterbalanced by a drop in very cold periods) is very likely associated with climate change; the increase in hydrometeorological episodes (intense rainy periods or droughts) and dynamic events such as storms or tropical cyclones is the object of a medium or weak degree of confidence. The other conviction rests on the probable increase in the frequency of climate hazards, an increase clearly associated with the level of concentration of GHGs reached and by the increase in global temperature over the course of the 21st Century. It is this conviction, supported by the models, which has led to the emergence in international policies of the notion of a "danger threshold" voiced since 1992 in the United Nations Convention on Climate Change, and which led, at least until the failure of the Copenhagen Conference (2009), to the establishment of objectives to limit the increase in global temperature to 2°C, and the concentration of CO_2 in the atmosphere to 450 ppm [MEI 09].

The link between the increase in concentrations of GHGs and the frequency or intensity of climatic risks is not however limited to a single threshold, however well-supported that might be. If some of the impacts are associated in a more or less linear fashion with the rise in global temperature, the majority of them are only manifested above a more or less raised level of temperature increase. The uncertainty over the probable effects of climate change is therefore growing, due to the increasing probability of ruptures, nonlinear relations or bifurcations within the climate system.

7.1.3. *Uncertainty: a key issue*

Uncertainty therefore emerges as a key element in the prospective approach associated with climate change. The succession of IPCC reports (1990, 1995, 2001, 2007) illustrates the growing role granted to the handling of uncertainty in our approach to climate change and its consequences. The term "uncertainty" in reality covers three very different realities:

– the first results from the limits or deficiencies of the methods or tools for simulating climate evolution. This aspect of uncertainty can be handled according to the usual scientific steps: model cross-comparison, sensitivity analyses, establishment of confidence intervals and criticism of results;

– the second lies in human societies' ability to limit climate change and to adapt to its effects. It is approached by the establishment of scenarios taking account of future evolutions in demography, the economy and international exchange systems and in particular, energy resource uses (and in particular fossil-forms of energy that produce GHGs);

– the last aspect of the uncertainty, without doubt the most important, refers to the current inadequacy of the knowledge needed to predict the climate system, such as socio-economic systems, and *a fortiori* their interactions.

In this context, and in spite of everything, experts try to provide policy-makers with results accompanied by an evaluation of uncertainty while leaving the latter with the responsibility of implementing policies for prevention or adaptation. Within the framework of preparation for the 5th IPCC report expected in 2013, the results put forward by the IPCC are thus accompanied by a "confidence level" to provide a nuanced view of the probability of expected climate phenomena occurring on a scale that goes from "extremely improbable" to "nearly certain".

The increased role of uncertainty is raised by the multiscalar nature of climate change: for example, the response in terms of a global macro-economy proposed by the "Stern report" [STE 07] seems to forget the diversity and complexity of the effects of climate change in time and space or the responses given to it and only appears to consider it across large global divides between developed economies, emerging countries and developing countries. Without discrediting the global approach to interactions between climate and economy, the putting into practice of concrete measures for reducing GHG emissions and for adaptation can only be conceived on the level of local areas, taking account of their specificities. It is from this perspective that the approach in terms of urban resilience acquires its legitimacy.

7.2. Adaptation to global change and resilience

Until the beginning of the 2000s, research on climate change developed an evaluation of its consequences based on the notion of impacts. Then, the term "adaptation" emerged, paving the way for resilience.

7.2.1. *Climate change: from impacts to adaptation*

The term "impact" reflects an approach based on simple causality. It was near-exclusive in the 1990 and 1995 reports (the only ones available at the time when the United Nations Convention Framework on Climate Change was negotiated – 1992 – and when Kyoto Protocol was negotiated – 1997). To simple causality, the approach added a sectorial vision (disruption to ecosystems, water resources, food security, rise in sea level and coastal risks, human health) which evidently left little space for interactions or for the systemic vision developed moreover in the management of risks. Broadly recognized by the modeling approach promoted by climate physics, this notion of impacts favors a forward-thinking response to the risks represented by the impacts of climate change: the reduction of GHG emissions, known by the term *mitigation*, thus became almost the sole way of limiting the magnitude of climate change and thereby its impacts.

During the 2000s, the difficulties of putting in place a policy for reducing emissions following the refusal of the federal administration of the United States and the extent of North-South divides prioritized the notions of vulnerability and adaptation to climate change. The geopolitical debate pits the Northern countries, considered responsible for climate change via their emissions (present and past), against the developing countries, assumed to be particularly vulnerable in the face of climate change. In this context, the IPCC reports dated 2001 and 2007 accord an important position to the evaluation of the vulnerability of areas faced with climate risk. Adaptation, defined as a process of reducing vulnerability, is presented as an aspect of development, a counterpart to the increasingly obvious limits of *mitigation,* understood here as the efforts agreed with a view to reducing GHG emissions.

In the international negotiations of the 2000s, the developed countries most reluctant to limit their efforts (the United States, Canada and Australia) became the most ardent defenders of a policy for adaptation. This has now been established by the announcement of an international Adaptation Fund after the failure of the Copenhagen Summit.

Outside of the geopolitical arena and the field of international negotiations, it is becoming increasingly obvious, even for nations committed to a policy for reducing their emissions, that the inevitable effects of climate change require national and local adaptation policies, as completely indispensable to *mitigation*.

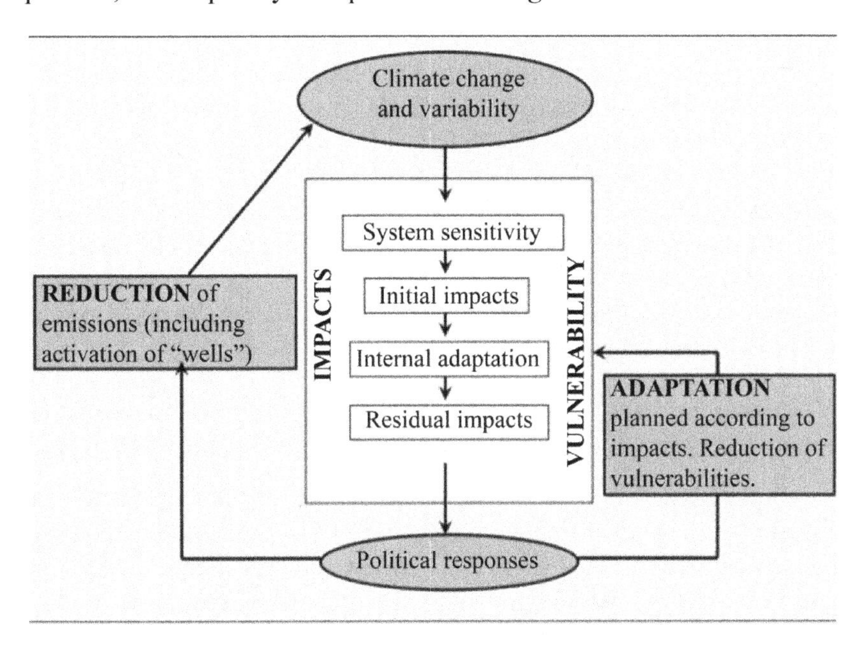

Figure 7.1. *Mitigation (reduction of emissions) and adaptation, two complementary aspects of the responses to climate change. From IPCC [IPC 07], adapted and translated by Claude Kergomard*

In France, the creation of the National Observatory for the Effects of Climate Warming (ONERC) and the implementation of national and local climate plans, which became Territorial Climate-Energy Plans, following the *Grenelle de l'Environnement*, thus mark the beginning of an era of adaptation to climate change. But this denomination of climate-energy plans also conveys the ever more direct association between adaptation in the strict sense of the word, that is to say the capacity for limiting vulnerability in the face of future climate events, and the objectives of the energy transition that creates the necessity of reducing fossil fuel use, as much due to the necessity of reducing GHG emissions as of their probable rarefaction.

7.2.2. *Adaptation to climate change: the diversity of themes and ambiguities*

The fusion, or confusion, between reduction of GHG emissions and adaptation to climate change, tends to prioritize the question of energy, indeed, even to make it almost the sole theme of publications and public policies dedicated to the sustainable city [MAN 11]. This is particularly the case in Europe, undoubtedly because of the European Union's commitment to the Kyoto protocol. This is evidently not the case everywhere in the world, for example for the large Asiatic cities included in the framework of the *Asian Cities Climate Change Resilience Network* [ACC 10].

Among the very broad range of questions raised in the ever more copious literature, four major themes can be distinguished [ROS 11]:

– the theme of energy, dominant in the developing countries engaged in policies for reducing GHG emissions and equally concerned with the security of their energy supplies in the foreseeable context of a reduction in nonrenewable resources; this theme interferes directly with urban transport policies and the habitat;

– the climate risks linked to extreme phenomena, evidently specific to each city depending on the particularities of the site and climate zone concerned: cyclones and storms, river flooding, etc. One particular case is that of large coastal urban areas, for which the rise in sea level is becoming a major aspect of climate risk;

– the security of water and food supplies, which very evidently concerns cities in arid and semi-arid developing countries;

– issues of public health caused directly or indirectly by climate change, such as the effects of heatwaves, but also the propagation of epidemic diseases "stimulated" directly by climate change, or, in big cities in developing countries, the difficulties of providing water or treating used water and waste disposal.

This evidently incomplete list suffices to show the inadequacy of a sectorial approach to urban adaptation to climate change. The lines between the themes are clearly arbitrary, and for many of the real events that can be linked, more or less exactly, to climate change and to the necessity for adaptation, interactions between themes are the rule and their interpretation requires a systemic approach. Thus, in France, the 2003 heatwave contributed a great deal to raising awareness of climatic risk, and drew the responses that form the Territorial Climate-Energy Plans, instituted by the National Plan for Adaptation and the laws resulting from the *Grenelle de l'Environnement* [GOD 11]. In slightly under two weeks at the beginning of August 2003, 15,000 people, mainly elderly people living in cities, succumbed to the heatwave. The excess mortality caused by the heatwave thus reached +134% in the Ile-de-France region, and exceeded +80% in a number of cities and urban areas with more than 100,000 inhabitants (Lyon, Bordeaux, Dijon, Le Mans, Poitiers) [INV 03]. Beyond the scandal that this death rate represented, the shortcomings in managing the crisis and the dysfunctionality in French society that it brought to light [GAU 03], the 2003 heatwave tested the necessary adaptation of the city itself to a specifically urban climatic risk, of which digital climate simulations for the second half of the 21st Century allow us to predict an increased frequency and intensity [PLA 08]. Indeed, among the factors that explain the particularly harmful effects of heatwaves on health, the phenomenon of a block of urban heat, directly linked to the shape and size of the city [OKE 73, OKE 08] figures in first place, and the common connection between high temperatures and episodes of urban atmospheric pollution. The capacity of mineral surfaces to store energy and to convert it into "sensitive heat", the radiative trapping caused by the "urban canyon" that vertical built surfaces create on either side of streets, the aerodynamic constraints exerted by buildings on air circulation which the urban atmosphere individual and as the abundance of primary pollutant production by motorized transport means pollutants are likely to be transformed into ozone during periods of strong sun and heat. For a long time associated with waste from industrial pollutants, atmospheric pollution has thus become a specifically urban phenomenon, directly associated with urban climatology.

The imperatives for adapting to climate change, for the limitation of GHG emissions and pollutants most harmful to health, are thus becoming more and more integrated into the urban prospective. But to associate, as the Territorial Climate-Energy Plans suggest, the reduction of energy consumption and emissions of GHG on the one hand, with the limitation of urban vulnerability in the face of climatic extremes on the other hand, can lead to objectives that are contradictory in some aspects. The increasing density of the city is often considered as a necessary objective for limiting the consumption of natural spaces, the increase in transport needs or the impermeability of soils and its hydrological consequences, but the increase in density can make blocks of urban heat more intense or increase biophysical vulnerability. Moreover, in the absence of real scientific certainties, the Territorial Climate-Energy Plans seem to be tools for raising the awareness of populations and mobilizing players in urban planning toward adaptation objectives, in the expectation that increases in knowledge and the lessons of experience will enable real solutions to be developed.

7.2.3. *Adaptation in question*

More broadly, well beyond the physical and material questions that the adaptation of cities to climate change poses, are other questions, both social and ethical, that are particularly pertinent in an urban environment. In a critique of strictly economic (cost/benefit) approaches to adaptation, the Stern report [STE 07] is emblematic; Adger *et al.* thus emphasize social factors likely to limit or favor adaptation:

> *We argue that, notwithstanding physical and ecological limits affecting natural systems, climate change adaptation is not only limited by such exogenous forces, but importantly by societal factors that could possibly be overcome* [ADG 09].

For these authors, the ethical values, the diffusion of knowledge within society, the nature of reports on society and inequalities, cultural positions in the face of natural risks and socioeconomic change are important elements in the adaptive capacity of a state or region.

The European project ESPON Climate [ESP 12] thus quantifies the adaptive capacity using the combination of a considered panel of indicators that integrates the economic resources of the region, the technological capacities, knowledge and diffusion of knowledge, infrastructure quality and the efficiency of institutions, since it is not able to consider explicitly the cultural and ethical values.

This approach thus combines, in its complexity, the definitions of resilience; it lends itself well to analysis on an urban scale, as carried out for example by *Asian Cities Climate Change Resilience Network* (ACCCRN), which defines resilience in the face of climate change:

> *Climate change resilience is the capacity of an urban community and institution to dynamically and effectively respond to shifting climate impact circumstances while continuing to function and prosper* [ACC 10].

Thus defined, urban resilience in the face of change therefore lies well-beyond effective practices of sustainable urban development.

7.3. Urban resilience and sustainable urban planning practices

For 15 years, at least in European cities, there is scarcely an urban development that does not refer to the objective of sustainable development, whatever the size or ambition of the project concerned. The question of adaptation to climate change is important.

7.3.1. *Temporal horizons and diversity in practices for sustainable urban development*

Taking account of the fact that transport (and in particular daily transport in built-up areas) and the habitat figure at the forefront of investments of energy economies, urban sustainability coincides to a large extent with the objective of creating the "post-carbon" city

[MAN 11]; in France particularly, reference to "Factor Four" as an objective to reduce the consumption of fossil fuel dominates the Territorial Climate-Energy Plans, which have become a compulsory reference for sustainable urban planning. More recently, adaptation to climate change, which includes the reduction of vulnerability in the face of extremes and climatic risks, management of the "urban metabolism" and management of health risks, are gradually being asserted, while still remaining subordinate to objectives for reducing energy consumption and GHG emissions [MAN 11].

As has been proposed by Theys [THE 09], it is possible to familiarize ourselves with the great diversity of practices on offer for sustainable urban planning depending on the temporal horizons in question and the degree of transformation in conceptions of urban planning implied:

– the first horizon is a short term horizon of normative or initiative measures likely to initiate progressive evolutions in the urban habitat and means of transport. Compatible with the shelf-life of national or local political mandates, this horizon includes, for example, the measures written into the framework of the laws of the *Grenelle de l'Environnement* in France, or urban developments aiming to increase provision of public transport and in parallel to discourage the use of individual vehicles in a number of built-up areas;

– the second horizon, lasting many decades (50 years), is at the same time one of large-scale urban planning projects and large-scale investments. This horizon corresponds to the time-period when climate change should be effectively sensitive. It is in this time-span that the realization of main objectives for transforming the urban fabric and ways of life needed to lead to objectives for sustainable development should be expected: a three- or four-fold reduction in energy consumption, a sensitive modification in schemes and behaviors in transport and a sensitive reduction in urban vulnerabilities in the face of climatic risks;

– the third horizon, by necessity poorly situated in a distant future, indeed relegated to Utopia, is that of the recomposition of urban forms and functions in line with major technological ruptures and a reconfiguration of political systems, ways of life and social relationships.

It is in the articulation of these different temporalities and the rationale of the players they underpin that a harmonious evolution toward sustainable development lies (the optimistic scenario) or the more or less harmful ruptures that put urban resilience in the face of climate change to the test.

7.3.2. *Eco-districts, ecological cities: what level of resilience?*

BedZed in London, the Vauban quarter in Fribourg, Kronberg in Hanover, Vesterbro, Hammarby Sjöstad and Bo01 in Copenhagen, Stockholm and Malmö, are some well-publicized examples of eco-quarters supposedly associated with extreme restraint in the use of fossil energies, indeed with a complete "carbon neutrality", a preponderance of public transport and/or green methods of transport, the climatic comfort caused by the presence of vegetation and water in the urban fabric, simultaneous with forms of social innovation in collective ways of life and local democracy. In parallel, there is scarcely a large built-up area in France or Europe that has not recently included in its major urban-planning projects one or several eco-districts taking account, to different degrees, of the principles of these pioneering examples that already date, for the most part, from several decades ago.

An experimenting ground for technological ("passive" buildings, use of renewable energies, green transport, etc.) or social innovations, emblematic representations of directions to follow toward the "sustainable city", the eco-districts symbolize an aspiration as much as they promote pragmatic solutions for limiting the negative impact of an urban spread that remains the predominant tendency. But criticism is not lacking, in particular over the social aspects of eco-districts, which the additional costs of technological innovation seem to reserve for the upper-middle classes: a social or generational exclusive tendency, exclusion from industrial activities, man-made environmental inequalities difficulties in sustaining the methods of community organization or government that generally accompanies the birth of eco-districts, etc. The existence of these quarters, of which it makes sense to give an instructive example [BON 10], only represents at most a few thousand homes, which only underlines the size of the task that the conversion of the existing urban fabric represents: the jump from the eco-district to the sustainable city is immense.

What can be said about ecological city projects, such as Dongtan on the edge of Shanghai or Masdar City in Abou Dhabi? These large-scale projects are sometimes presented as a future solution to the evils associated with rapid urban growth in these countries. Their effective implementation still remains very hypothetical and can only, in any case, offer very limited perspectives in the face of the challenges that global urban growth represents. The example of Dongtan is particularly demonstrative: Dongtan is an "eco-city" project likely to welcome 25,000 inhabitants from 2010, and several hundred thousand toward the middle of the 21st Century. It has been cited in a number of articles as "the first ecological city in the world" [LAN 06]. On a site of 8,400 ha (around 2/3 the surface area of Manhattan) made up of alluvial sediments dammed up for 30 years to the east of the island of Chongmin, in the Yangzi Delta to the north of Shanghai, the project foresaw a city, made up of three village groups, distinct but linked to one another, entirely autonomous through their production of renewable energies, practicing the recycling integral to their water consumption and waste, abolishing the use of transport consuming fossil energies and favoring "green" transport. The ecological imprint was expected to be more than twice less than that of Shanghai. Around the city, the polders would be converted into biological agriculture, and the natural reserve of Dongtan, a vast coastal wetland listed in the inventory of the Ramsar Convention, would form its natural horizon. Scarcely three years after the sensational announcement of the project, it already seems dormant. Dongtan will be without doubt the definitive "*eco-city that never was*" [WIL 09]. The completion, in October 2009, of a 25 km bridge-tunnel complex that links on the one side the Yangzi estuary to the Isle of Chongmin, to the rest of the city of Shanghaï does not encourage optimism. During a visit to the island of Chongmin in March 2011, we were able to state that its future lies without doubt in the integration of the island with the suburban space of Shanghai. On the planned site for the eco-city, a network of out-sized routes serves some developments in the process of completion: residences destined for a yuppie clientele of Shanghaians or foreigners, to whom the bill-boards and publicity leaflets boast the environment quality (sea air, proximity to the coast and the nature reserve) but which do not correspond in any way to the initial objectives. Only the presence of some wind turbines and an organic farm in the polders recalls the initial project. Elsewhere on the island, the accelerated destruction of the rural habitat and the gathering of the population into large collective housing complexes is supported. The Dongtan nature reserve, subject to pollution by heavy metals and colonization by an invasive species, *Spartina alterniflora*, is witnessing a rapid decrease in bird populations. This somber tableau raises questions about the reasons behind the failure of a project that had aroused so much passion and about the resilience of a sadly ordinary form of urban expansion [OBR 09]: the political-administrative context in China, real-estate interest and the effects of corruption (the Mayor of Shanghai, an active project developer, was dismissed for corruption at the end of 2006) and the interest of a publicity operation on a global scale for the major London planning-firm involved [OBR 09].

Figure 7.2. *On the site of the "eco-city", Dongtan. Photography: Claude Kergomard, March 2011*

The example of Dongtan illustrates extremely well the risks attached to ecological urban development *ex nihilo*, regardless of any political, economic or sociological context, and with little forethought, which link urban resilience and the adaptation of cities to climate change.

7.4. Conclusion

Beyond the technical solutions that eco-districts or "carbon-free" city projects represent more or less realistically, adaptation to climate change calls into question the economic, social and political forms of urban development.

As Peyrache-Gadeau and Pecqueur [PEY 11] remind us, urban development is still generally considered to be the result of the concentration of populations, industry and commercial activities, which themselves generate specific resources (expertise, technology, cultural heritage, etc.), which perpetuate and prolong the urbanization and metropolization process. But this concentration is also the cause of harm and risks ("urban spread", pollution, vulnerabilities and inequalities) which challenge the "modern" representation of an invulnerable city in continuous development. The adaptation of cities to climate change in its broadest sense is inconceivable without mobilizing resources and innovative capacity in all its forms; it should also take account of the uncertainties attached to climate change and the "surprises" that can be encountered in the context of an energy transition, of which the scenarios remain multiple and very hypothetical.

In this regard, a city resilient in the face of climate change and a future of uncertain energy supplies meets the definition of the sustainable city put forward by Emelianoff [EME 02a, EME 02b]: a city capable of mobilizing technological, social and cultural resources, within the framework of a collective project promoted by a well-adapted government.

7.5. Bibliography

[ACC 10] ACCCRN, Introduction to ACCCRN, available at: www.acccrn.net/resources/introduction-acccrn-prepared-bellagio-donors-meeting, 2010.

[ADG 09] ADGER W.N., DESSAI S., GOULDEN M. *et al.*, "Are there social limits to adaptation to climate change?", *Climatic Change*, vol. 93, pp. 335–354, 2009.

[BOM 10] BONARD Y., MATTHEY L., "Les éco-quartiers: laboratoires de la ville durable", *Cybergeo*, available at http://cybergeo.revues.org/23202, 2010.

[BRA 09] BRACONNOT P., DUFRESNE J.-L., SALAS Y MELIA D. *et al.* (eds), Analyse et modélisation du changement climatique, 2nd ed, Livre blanc Escrime, Société Météorologique de France and Météo-France, 2009.

[EME 02a] EMELIANOFF C., "Comment définir une ville durable?", available at http://www.ecologie.gouv.fr/IMG/agenda21/intro/emelia.htm, 2002.

[EME 02b] EMELIANOFF C., "L'urbanisme durable en Europe, à quel prix?", *Écologie & Politique*, no. 29, pp. 21–36, 2002.

[ESP 12] ESPON CLIMATE, "Climate change and territorial effects on regions and local economies in Europe", Internal report, 2012.

[GAU 03] GAUDIN J.-P., "Une canicule politique", *Cybergeo: European Journal of Geography*, available at http://cybergeo.revues.org/5404, 2003.

[GOD 11] GODINOT S., 2011: "Les plans climat énergie territoriaux: voies d'appropriation du facteur 4 par les collectivités et les acteurs locaux?", *Développement durable et territories,* vol. 2, no. 1, March 2011. Available at http://developpementdurable.revues.org/8874.

[INV 03] INVS Impact sanitaire de la vague de chaleur d'août 2003 en France Bilan et perspectives, available at http://www.invs.sante.fr/publications/2003/bilan_chaleur_1103/vf_invs_canicule.pdf, 2003.

[IPC 07] PARRY M.L., CANZIANI O.F., PALUTIKOF J.P. *et al.* (eds), "Contribution of Working Group II to the Fourth Assessment Report of the Intergovernmental Panel on Climate Change", *in IPCC, Climate change 2007: impacts, adaptation and vulnerability,* Cambridge University Press, Cambridge, UK, available at http://www.ipcc.ch/, 2007.

[LAN 06] LANGELLIER J.-P., PEDROLETTI B., "La première ville écologique sera chinoise", *Le Monde*, 16 April 2006.

[LEI 11] LEICHENKO R., "Climate change and urban resilience", *Current Opinion in Environmental Sustainability*, vol. 3, no. 3, pp. 164–168, 2011.

[NEW 09] NEWMAN P., BRADLEY T., BOYER H., *Resilient Cities, Responding to Peak Oil and Climate Change*, Island Press, 2009.

[MAN 11] MANCEBO F., "La ville durable est-elle soluble dans le changement climatique?", *Urban Environment*, vol. 5, available at http://www.vrm.ca/EUUE/vol5_2011/EUE5_Mancebo.pdf, 2011

[MEI 09] MEINSHAUSEN M., MEINSHAUSEN N., HARE W. *et al.*, "Greenhouse-gas emission targets for limiting global warming to 2°C", *Nature*, vol. 458, pp. 1158–1162, available at http://www.nature.com/nature/journal/v458/n7242/abs/nature08017.html, 2009.

[OBR 09] OBRINGER F., "Les écovilles en Chine: du rêve nécessaire à la réalité marchande", *Mouvements*, vol. 4, no. 60, pp. 29–36, 2009.

[OKE 73] OKE T.R., "City size and the urban heat island", *Atmospheric Environment*, vol. 7, no. 8, pp. 769–779, 1973.

[OKE 08] OKE T.R., "Street design and urban canopy layer climate", *Energy and Buildings*, vol. 11, nos. 1–3, pp. 103–113, 2008.

[PEY 11] PEYRACHE-GADEAU V., PECQUEUR B., "Villes durables et changement climatique: quelques enjeux sur le renouvellement des 'ressources urbaines'", *Urban Environment*, vol 5, 2011, available at http://www.vrm.ca/EUUE/vol5_2011/EUE5_Peyrache_Gadeau_Pecqueur.pdf, 2011.

[PLA 08] PLANTON S., DÉQUÉ M., CHAUVIN F. *et al.*, "Expected impacts of climate change on extreme climate events, *Comptes Rendus Geoscience*, vol. 340, nos. 9–10, pp. 564–574, 2008.

[ROS 11] ROSENZWEIG C., SOLECKI W.D., HAMMER S.A. *et al.* (eds), *Climate Change and Cities: First Assessment Report of the Urban Climate Change Research Network*, Cambridge University Press, 2011.

[STE 07] STERN N. *et al.*, *The Economics of Climate Change: The Stern Review*, Cambridge University Press, 2007.

[THE 09] THEYS J., "Scenarios pour une ville post-carbone", *Constructif*, no. 23, July 2009.

[WIL 09] WILLIAMS A., "Dongtan: the eco-city that never was", *Spiked*, available at http://www.spiked-online.com/site/article/7330/, September 2009.

8

Organizational Resilience: Preparing and Overcoming Crisis

Resilience can be understood as the ability to overcome a crisis. This definition comprises resilience in both its reactive and proactive dimensions (respectively, the ability to withstand disruption and the ability to increase one's capacities for self-organization and learning) [DOV 92]. These are considered here with regard to the result (was the system capable of absorbing the shock?) as much as with regard to capacities in a strict sense (are the systems' adaptive properties, acquired before the crisis, used during it?).

Overcoming a crisis means guaranteeing an organization's continued functioning, even in degraded mode, in order to facilitate a return to normal as rapidly as possible. This involves relying simultaneously on different competencies that touch on issues of organization and anticipation, as much as on the ability to withstand stress or even issues of co-operation between players. Recent disasters have demonstrated both the multidimensional nature of some crises and the necessity of developing strategies for mitigation, preparation and adaptation on a collective scale as much as on an individual scale.

Strategies for mitigation and preparation involve considering the import of the risk culture as much as the factor of resilience. Indeed, an event of the same nature and gravity can be felt as an individual or

---

Chapter written by Richard LAGANIER.

collective shock, or even as a single episode in a personal or collective trajectory. The difference depends largely on the quality of organization and learning that a society or an element of it (an enterprise and an individual) has developed for managing or predicting crisis.

This condition is essential for resilience and is determined by a combination of diverse variables, which may be: material (communication networks, and resources allocated for producing and circulating information in times of crisis), structural (leadership, coordinating multiple levels of action, etc.), organizational (preparation, prediction and involving many different crisis scenarios), psychological or moral (education, values and confidence).

8.1. The components and temporalities of a crisis

Constructing a plan of action for managing a crisis and entering into a learning process calls for a return to the definition of what makes a crisis in an operational perspective.

8.1.1. *System disruption*

From the 1990s, P. Lagadec distinguished emergency situations from crisis situations. An emergency is defined as a known event, for which codified procedures are available, involving a limited number of players, integrated into a single, clearly defined authority structure [LAG 93]. A crisis means an event that produces a profound instability in the system and is not covered by the usual response frameworks, undermining security or public safety or even the integrity of government structures. This distinction fits the analyses carried out by G.Y. Kervern on the development of cindynics [KER 95], which distinguishes events according to the degree of disruption they cause to social organization.

Gaps at management level in an emergency situation, whatever its cause (a natural disaster, an epidemic, an accident, an act of terrorism, a cyber-attack, a breakdown and major faults in information systems, and disruption to public services), can thus produce a crisis. Because of its disruptive effects (change in an organization's tasks, rules or indeed values), a crisis requires a rapid response at the same time as the references and operational frameworks for action are dissolving, frameworks that allow guidance and supervision and give sense and value to individual and collective action in the context of an emergency.

8.1.2. *The components of a crisis*

The deciphering of recent crises underlines the presence of several recurring characteristics. The explosion at the AZF factory, on 21st September 2001 in Toulouse, for example, revealed the inability of care systems to absorb a rapid influx of wounded a few minutes after the initial shock and showed the weaknesses of the contingency plan, which anticipated that the factory would seal itself off in case of chemical hazards, when the blast from the accident had destroyed the windows. The heatwave that affected Europe in 2003 and triggered 15,000 deaths in France over 15 days, similarly revealed the public authorities' lack of foresight and a profound lack of co-operation between individuals, as well as a poor knowledge of the reflex actions for overcoming this type of situation.

P. Lagadec notes that crises often develop through a flood of problems that exceed response capacities as they have been envisaged. The size and speed of the event render emergency logistics useless. Protection systems are shown to be inappropriate. They can even, as a result of their failure and the image of security that they have been able to convey in the past (protective dykes and industries very much controlled by independent agencies), cause major damage in spaces considered to be "safe", and indeed beyond, by a diffusion and contagion effect. Moreover, the interplay of dependencies between components of the system, the intertwined nature of essential infrastructures (energy, water, transport, communications, etc.), the complexity and the random nature of some crises renders the outcome of decisions taken uncertain and many of the emergency plans initially envisaged ineffective. Moreover, the systems ensuring crisis management can themselves be disrupted. Thus, during the rain storm of 3rd October 1988 (between 300 and 420 mm fell over a timespan of 4 and 12 h), which triggered severe flooding in the city of Nîmes, most crisis management infrastructure, although accustomed to managing emergencies, was unusable because it had been flooded.

Operational structures can also be disrupted by latent conflicts or discord between stakeholders usually united in a crisis situation. This was the case during the flooding of the Somme, where the local and national representatives, on the basis of an election and a rumor (that Paris had flooded the Somme by diverting the water, in order to prevent flooding in the capital), clashed on the methods used to manage the crisis as well as on the speed with which the authorities in power reacted to confront the crisis.

In sum, a crisis always disturbs a socioterritorial system in its entirety, via its components. It threatens priorities and challenges traditional behaviors and shared values within a social organization (a social group, enterprise, etc.) [STA 03]. Crises have impacts on individuals, property, different kinds of heritage and brand image, and thus threaten the very survival of the organization they affect. They can constitute, through their multiple consequences, a factor for unbalancing, breaking down and profoundly disintegrating systems, to the point of leading later on to a change of state in the affected system.

8.1.3. *The time of crisis*

A crisis makes itself felt depending on several temporalities. First, it often emerges in an environment that is favorable to it. Indeed, an incubation period, of variable duration, builds up different forms of latent vulnerabilities (organizational, technical, cultural and social weaknesses). From this point, the crisis develops following a trigger event of natural or man-made origin, which reveals an environment or organization weakened by the presence of these underlying vulnerabilities. This trigger event can be single (a flood, an act of terrorism, etc.) or multiple (the domino effect during the recent events in Japan combining an earthquake, a tsunami and industrial accidents), unforeseeable, often with a low probability of occurring and great intensity. This phenomenon affects the system severely, indeed destructively, triggering a state of shock and surprise among populations and emergency managers so strong that its gravity was unthought of during planning.

A crisis should not, therefore, be considered as being exogenous to organizations. It is only the revealer of multiple vulnerabilities that existed before it was triggered. The creation of the crisis does not moreover coincide with the appearance of the trigger event. It emerges and prospers in reality on the fertile ground of organizations' lack of adaption to extreme phenomena and a lack of foresight.

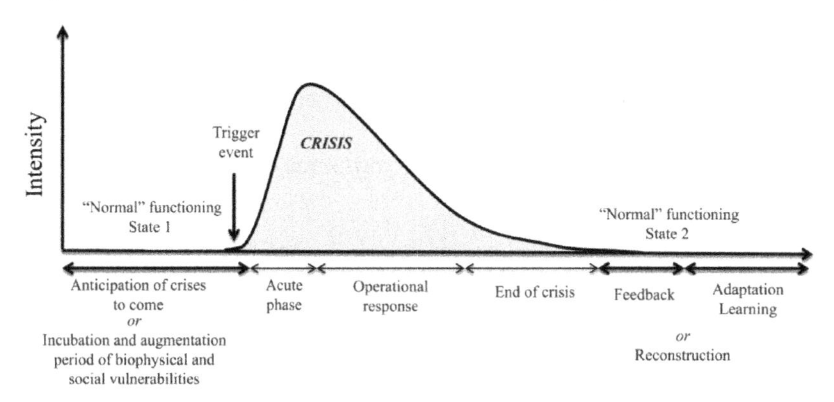

Figure 8.1. *The different phases of a crisis*

Crises begin when organizations lose control of the trigger event and impacts. Aggravating factors (rumor, panic, domino effects, emergency structures' inability to intervene immediately, etc.) can amplify the initial situation by making management and resolution of the situation more complex. They then die down gradually, in line with the operational responses that are applied to end the crisis.

8.2. Lessons from feedback

After a crisis, it is becoming increasingly common to collect experience feedback. This enables lessons to be learnt from past errors, faults to be detected and a learning process to be commenced as well as a readjustment of the means of managing crises and reducing vulnerabilities.

Feedback aims to retrace in detail the evolution of a past event and the sum of actions taken to overcome it, and then to analyze this information to identify beneficial steps. It is generally carried out within an organization (a business, a local community, public services, etc.) [WYB 09]. It can be encouraged or made obligatory by an organization's executive, particularly in an accident or crisis.

The implementation of experience feedback seeks to meet five objectives:

– to keep a record of crisis situations, especially in order to take statistics into account;

– to help correct anomalies in performance (technical breakdowns and poor decisions);

– to share knowledge of events in order to make collective progress in the field of activity;

– to identify scenarios at the root of potential crises;

– to bring positive aspects of crisis management to light in order to promote them later on.

From this perspective, an experience feedback seeks to gather and analyze different types of information on the event and its management: general information (the date of the event, its duration and location); the typology of the trigger event (flood, fire, explosion, etc.) as well as its origin; accompanying context; players implicated in the event and its management (role, and responsibility for acts and decisions); a chronology of acts and decisions over the course of the event, from the period preceding the emergency situation until normality has been restored (several weeks after the crisis); the consequences of the event (human impacts/impacts on property or the environment, and media reports).

This feedback, discussed with players in the crisis, enables the biggest areas of weakness and failure to be identified and ways for improving organizations and making regions more resilient to be found. In particular, this concerns regulations for prevention, protection or emergency management. If some of the recommendations can be adopted in the handling of an emergency itself, others should be integrated into more long-term planning (the introduction of new tools for action and modification of existing public action).

8.2.1. *Managerial flaws*

Due to experience feedback, numerous players have highlighted the most common failures in crisis management [LAG 93, KER 95, FIN 96, UN 05].

Failures in warning and emergency preparation are frequently cited among the weakest points: a lack of simulation and preparation exercises for the crisis, a lack of management plans, the absence or inefficiency of the crisis unit and the weakness of warning systems. The absence of preparatory simulations can thus lead to stress for decision makers and destabilization of the teams that have to face extreme situations. This results in decisions and behaviors that may appear irrational after the event: an absence of communication, a refusal to take up responsibilities, denial of the event itself, a feeling of invincibility or on the contrary, deep-seated panic.

Numerous failures also appear in managing the acute phases of crises: crisis teams transformed into a “bunker” mentality removed from reality, contributing to situations being underestimated and solutions being poorly adapted; those in positions of permanent responsibility being unreachable, whereas they should ensure the coordination of activity on the ground and inform the crisis unit of changes in the situation; crisis handling is implemented too late due to an absence of decision-making and lack of reactivity; an absence of internal communication, inadequate communication between players (transmission of inaccurate information, events being minimized and the failure of communication technologies); insufficient resources allocated – both human as well as logistical – which makes the management system less effective or indeed inoperable; and emergency procedures "put out of action" as a result of the violence or speed of the event.

Weaknesses in response to immediate post-crisis phases have also been identified. This involves in particular the absence of feedback or the failure to take into account recommendations specified and explained by experience feedback.

8.2.2. *Cultural deficiencies*

Many deficiencies of a cultural kind can also come to weaken organizations poorly prepared for crisis:

– a culture of infallibility also leads to errors of judgment and ill-advised behavior: for instance, the supposed unsinkability of the Titanic was disproved on 14th April 1912; the strength of nuclear power plants in modern, hyper-technological Japan, reputed to be well prepared for crises, was discredited on 11th March 2011 following an earthquake of magnitude 9 and above all the tsunami that followed it;

– the culture of simplification and a denial of complexity can translate into poor anticipation and inappropriate handling of danger. Thus, the domino effect illustrated by the Japanese nuclear disaster had not been anticipated.

In other events, deficiencies in communication between social groups, just like deficiencies linked to linguistic barriers between countries or differences in professional cultures, can cause delays or ruptures in the information, or indeed disinformation transmission chain and worsen a crisis. Menoni [MEN 01] points, for example, to the "rigidity of organization in the Japanese system" during the Kobe earthquake in 1995 and the difficulties in communication between civil and military players. In the event of crisis, those responsible had to conform to a very strict system of codes and procedures, which led to particularly prejudicial delays in the transmission of information and the deployment of aid. To mobilize the military personnel, a written request from the governor was mandatory, detailing the number of men, boats and airplanes required. The governor had moreover to produce a precise report on the situation and the devastated areas to obtain this military support, which, to use S. Menoni's terms, was unreasonable given the extent of the damage and the confusion that reigned.

Finally, it appears that crisis is a powerful disruptive factor since it undermines exactly those cultural foundations that usually guide behavior. If we use the example of Kobe, S. Menoni emphasizes that:

> *in exceptionally stressful conditions, as psychologists working with Kobe earthquake's victims showed, decision makers and people in charge of emergency operations fail to understand that changes may occur in normally shared values and habits, resulting in a gap between people's expectations and bureaucratic norms adopted for facing the disaster. This conflict is worsened by the many uncertainties involved in any decision to be taken under emergency conditions* [MEN 01].

8.2.3. *Organizational failings*

Many examples, especially during crises linked to industrial accidents, have moreover clearly shown the predominance of the productivist system and the benefit of aspects of safety and security.

On a completely different note, the case of contaminated blood proves how diluting responsibilities can lead to a lack of even essential decisions in anticipating and stopping the emergence of a crisis as early as possible. Major cross-border crises, such as the accident at the Sandoz factory near Bale in Switzerland on 1st November 1986, the cause of major pollution in the Rhine or the accident at the nuclear power station at Chernobyl on 26th April 1986, have also shown the difficulty of acting in a coordinated manner on an international scale. Environmental protection and the safety of property and people indeed come into play across national boundaries and national considerations. Additionally, deciders should integrate the international dimension of this type of crisis by building on the development of international networks [ANS 10]. The latter can contribute to exchanges of information and knowledge, to the definition of agreements, norms and common objectives, the coordination of national initiatives and the reinforcement of warning and monitoring systems, especially to face new risks or little known hazards (pandemics, terrorism, tsunamis, etc.).

The multiple deficiencies identified via feedback demonstrate, if necessary, that the response to be developed to overcome crises cannot be confined to handling a specific event. Crises, whatever the trigger event, in fact affect the system globally and polymorphically. They belong in the context of growing complexity in intervention frameworks linked to multiplying levels of reflection and action, to the increase in the number and variety of instruments to be used to anticipate, mitigate or manage crises, and the emergence of new players in decision-making processes, players not always used to working together.

8.3. Organizing to overcome a crisis

It is in this context that the new spatial and scalar configurations for governing crises are defined. They result from processes that are judicial or rule-making in origin, from regulations, prediction, adaptation and learning-processes that bridge different levels of intervention (from international to local).

Henceforth, crises management is based on an integrated approach that aims to combine the three temporalities of crisis (before, during and after) and to take account of increasing complexity due to the interconnectedness of levels and players and the interdependence of processes that engender and accentuate crises. Henceforth, organizing to overcome a crisis lies in processes and actions for mitigating and preparing before the crisis, a reinforcement of intervention methods and structures to address the consequences of crises and guarantee intelligent communication during the course of crises and processes of re-establishment [BOI 03].

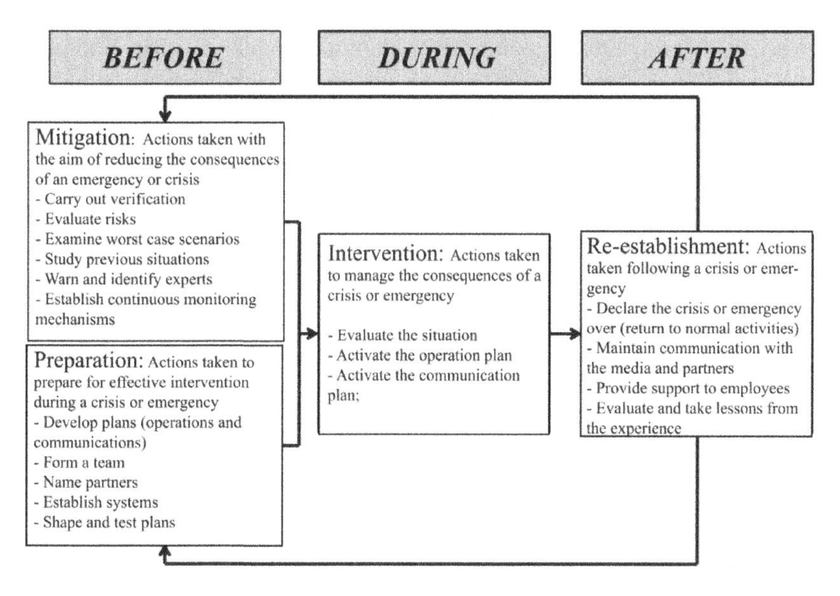

Figure 8.2. *The process of managing crises and emergency situations (from [BOI 03])*

8.3.1. *Before: mitigating and preparing for future crises*

Actions to be followed before a crisis consist of two types of complementary approaches: mitigation and preparation.

Actions for mitigation aim as a priority to identify and anticipate vulnerabilities in a system faced with particular types and levels of hazard. Identifying points of physical weakness enables some elements of the system to be protected in order to limit the impacts on the latter in case of crisis. This approach also relies on expertise that aims to identify and evaluate critical situations. It is based on the examination of imaginable scenarios and on the study of earlier crises or emergencies. It seeks in fact to consider the expertise required for likely crises or emergencies, and to establish mechanisms for continuous monitoring and warning.

There are numerous examples to illustrate earlier approaches already used for mitigation. When tackling floods, it is crucial to simultaneously carry out actions raising awareness among the actors involved, to take into account any potential danger in urban planning or mitigation documents as well as taking actions towards reducing vulnerability, monitoring, warning and providing protection when faced with hazards.

Areas for action aiming to reduce the risk of flooding	Types of action
Information and communication	*Atlas des Zones Inondables* (Atlas of zones at risk of flooding); *Document d'Information Communal sur les Risques Majeurs* – DICRIM (Common Information Document on Major Risks); Flood marks; disclosure requirements for solicitors and estate agents in property transactions
A collective consideration of the risk of flooding during land development	Forbidding construction in flood zones or making construction conditional on permission: *Plan de prévention des risques inondation* (PPRi) (Flood Risk Prevention Plan) appended to the local urban development plan, *Schéma de cohérence territoriale* (Scot) (Territorial Coherence Plan)
Reduction of biophysical vulnerability on an individual scale	Carrying out diagnostics involves; financing and carrying out work to adapt buildings to zones prone to flooding; to store items of value safely, – these are often light and easy to transport; to store vehicles safely by removing them from garages or car ports and by parking them on the nearest high ground (already located); and to store polluting materials and products safely
Protection	Collective: maintenance and monitoring of protection works (dykes, barrages, flood barriers, etc.); creation and maintenance of flood expansion zones, overflow outlets, storm basins, etc.

	Individual: installation of flood reservoirs in front of doors; relocation, elevation of buildings, waterproofing of equipment and networks, in particular of electronic systems (circuit breakers, switches, cables, fuses, etc.)
Monitoring and warning	Implementation of 22 flood forecasting services, *services de prévision des crues* (SPC) in France and the *Service central d'hydrométéorologie et d'appui à la prévision des inondations* (SCHAPI) (central hydrometeorological service and support for flood forecasting), based in Toulouse close to the main Météo France services, to ensure coordination of flood forecasting at a national level; personalized warnings, telephone warnings at home

Table 8.1. *Possible steps for reducing flood risk*

As for preparation activities, they aim to set out intervention parameters, administrative methods, resources, training and the testing of operational intervention plans in crisis and post-crisis situations, as well as communication plans. They are accompanied by collective or individual learning steps that aim to reinforce the ability of directors as well as the public to overcome extreme situations. In particular, it is a question of acquiring the right tools (crisis management manuals, response sheets, and communication and coordination plans) and putting in place a network of players involved in control, informational and operational roles on strategic (executive) and operational levels (crisis teams) as much as tactical ones (forces deployable on the ground). These activities thus seek to identify and train players to be mobilized in periods of crisis (internal players, partners, institutions and media) and to conduct simulation exercises to test the viability of the action plan.

On an operational level, it will then be a question of identifying crisis and emergency situations as well as possible in order to define steps whose objective will be to reduce the impact at the moment of crisis. The challenge lies particularly in implementing steps to improve interaction between personnel, in order to enable organizations to become more robust and to try collectively to overcome situations that are unforeseeable or outside of normal experience. In particular, endeavors to improve internal or external direction on levels involved in crisis management (ground teams, crisis teams and executive) are needed.

This type of operational plan also seeks to define administrative procedures and to identify those responsible for activating its use. It defines procedures for streamlining information circulation between operators on the ground and the crisis team, thus facilitating decision-making. Indeed, it defines which installations are adequate system supports for crisis management (operation centers and logistical needs with regard to infrastructure, human and technological resources).

Moreover, public authorities just as much as businesses should develop action plans to reduce the impacts of disasters, to ensure functioning in degraded modes or the transfer of activities or indeed the elimination of some vulnerabilities, in particular those related to infrastructure essential to the functioning of the urban environment (electrical and transport networks; hospitals, etc). These different questions necessitate significant work between services with a view to promoting a shared risk culture and creating a co-ordinated plan of action, which considers the interdependence of services and intervention levels in crisis periods. For example, businesses mobilize major operating departments as well as support functions such as security. But, taking into account the complexity of some crises which cannot be managed by a business alone, the response should also be developed in conjunction with public authorities (anticipatory work, aiming, for example, to share out procedures or to carry out simulation exercises jointly).

Thus, in France, each level of responsibility and competence has a crisis management tool for leading and managing inter-ministerial stakeholders. At local level, this is Departmental Operational Center (DOC), at zonal level, this is the Center for Zonal Operations (CZO) and at national level, the Center for crisis management in the ministry in charge of the sector concerned, even the Inter-ministerial crisis center (ICC) for large-scale crises requiring inter-ministerial coordination. At local and zonal level, DOC and CZO integrate the leadership of the main decentralized State services (representative(s) of the Regional health agency, representative(s) of local education authorities, army representative(s), representative of regional state offices (environment, development and housing), etc.). At national level, ICC defines the action of crisis centers in main government departments covered by the Interior Ministry (COGIC for civil security, COP for the police and CROGEND for the *gendarmerie*), as well as other ministries (environment, health, agriculture, industry, etc.).

At local levels, the Commune Safeguard Plans (CSPs) [GRA 08] have been developed to ensure that communes can prepare and organize themselves in the face of emergencies. Some 5,000 communes out of more than 36,000 have now signed up to it. The setting of the CSP is based on a diagnostic of risks. It is also responsible for warning and informing the public, ensuring that public and private means for responding to emergencies are listed, enables the creation of the crisis organization which will have to interact, if the case arises, with other intervention levels, suggests practical tools (reflects on the pragmatic questions, "who does what, how?") and requires the project to be sustainable over time (by updating exercises and procedures). Several success factors for the CSP and the durability of its operational character should also be taken into account. This involves in particular foreseeing the participation of the maximum number of individuals (politicians, agents, etc.) in elaborating it to promote its efficiency and use by those responsible. It is sensible here to devise simple but efficient tools rather than a volume that is not usable for those involved. Finally, it seems necessary to put in place regular exercises in order to test all or part of the CSP and to instill a principle of continuous improvement.

This operational approach should be combined with efficient communication and appropriate plans. It aims especially to provide the strategic and tactical manuals necessary for good communication in crises. The challenge is to enable all those involved to have the same benchmarks and the same reactions during the operation. It should also identify the named spokespersons who will have to liaise with the media, identify target audiences (internal and external to the organization or region) in order to maintain a permanent link with the individuals concerned, to specify facilities and training needs, etc.

Preparing better also means developing training tools for crisis management as P. Lagadec suggests, in particular because some crises not included in known frameworks do not fit the predefined patterns and action plans that could have been anticipated [LAG 93]. Among these tools figures the organization of seminars with the senior officials responsible for handling general leadership problems and communication in situations of extreme disruption. Carrying out audits of the vulnerabilities and potential weaknesses in a region also enables the potential risks to a business or urban network to be identified or indeed reduced. Thus, plans can be developed for the continuity of activity or plans for functioning in degraded mode adapted to known vulnerabilities to better overcome future crises.

C. Roux-Dufort has thus identified four types of post-crisis learning taken from feedback:

– to decide there is "nothing to learn" and to continue as if the crisis had merely been a one-off event in the life of the organization";

– "to benefit from the crisis to improve the present";

– "to benefit from the crisis to question the assumptions of an inadequate management";

– "to benefit from the crisis to redefine a business's identity" [ROU 00].

If feedback enables lessons to be drawn from past crises, the evaluation of response systems already at work also invites a general questioning of emergency plans and the deeper ability to become familiar with vulnerabilities and dependencies with regard to elements external to the system. It contributes to continuously reinforcing practices and expertise. For this reason, simulation exercises can be used for the conception and evolution of response systems. Carried out periodically, usually with the support of external experts, they enable essential lessons to be drawn to improve systems' reactivity and efficiency. Finally, training for specific players (a crisis director, the head of a business, the DGS and senior officers, and services involved in crises such as security teams) can complete this shared learning plan: beyond crisis management, preparing directors involves training them to react to complex situations, not included in the established plans, so as to lead them to develop strategic and organizational capacities that enable severe disruption to be withstood.

Finally, learning involves the average citizen. Crisis preparation aspires to make the individual an actor in their own security, which underlines the law to modernize civil security of August 13th 2004, which specifies that "Civil security is a collective issue. Each citizen contributes to it by their behavior. A proper culture of preparation for risk and threat should be developed." This law aims to prepare individuals to behave well and to react reflexively and raises awareness of the role of emergency services for establishing confidence in these organizations to reduce hitches in emergency operations. This preparation seems very effective when directed at children and teenagers. Consequently, training aiming to raise public awareness of the risk of earthquakes or hurricanes is carried out in schools in some countries or in France in the most exposed areas (and especially abroad, to make pupils aware of how to behave in the event of a hurricane or earthquake). Information sessions at the beginning of the school year on schools' responsibility for pupils during hydrological crises are now scheduled increasingly, for example, in order to limit risk-taking by parents determined to take to the road to look for their children when danger is announced [RUI 04].

8.3.2. *During the crisis: intervening to limit impacts and control the situation*

While the crisis still occurring, the aim is at once to limit damage and control the situation. Intervening in a disturbed situation, indeed in a chaos situation, demands a complex level of operations and communications coordination. This varies, however, depending on the nature and magnitude of the crisis.

The first stage is to evaluate the situation: to collect all the information describing the situation, while verifying the quality and reliability of data in order to assess the importance and nature of the problem; to confirm information from witnesses using objective details obtained from different sources, and in particular from emergency service staff who are on site (fire brigade, gendarmerie and police); and to map the site affected by dispatching an airplane or helicopter there. The rapid transmission of these data should be anticipated.

Enabling intervention and communication plans to function in case of chaos is one of the major difficulties of organizing intervention in situations that lie outside classic management frameworks: preparing a reliable and robust information system facilitating exchanges of information and the influx of communication in spite of the much degraded context is an essential challenge for the efficiency of crisis teams.

Once appraised of the situation, the decision maker (prefect and minister) determines the structure responsible for intervention (crisis team). This then activates the operation plan and the communication plan. The crisis team is responsible for directing interventions on the ground, by sending first aid teams with their equipment (medical assistance and support for firefighters) as a priority and if necessary re-establishing telecommunications (equipment for mobile telephones at least); organizing means of assistance for those affected by ensuring they have a food supply and shelter, even if it is very basic (e.g. tents), means of heating (in winter) and lighting.

At the same time, the communication structure aims to set out the official position of those responsible for the crisis team. All authors who have worked on crisis communication note that in this respect it is essential to take the initiative and never to be on the defensive, to communicate errors and lacunae and only to communicate known and exact information while remaining open and transparent. The task will consist of recognizing and explaining the nature of known reasons for the problem, and explaining the immediate measures deployed to control the situation. After the event, an official declaration that the crisis is over will be made.

8.3.3. *After the crisis: re-establishment*

Once the crisis has passed, a number of actions remain to be taken so that the districts and players affected can be re-established as quickly as possible. The care of victims following psychological shock, devastation, stress and fatigue is essential. Overcoming degraded living conditions and difficulties returning to normal (tackling insurance companies' and administrators' red tape) requires support to prevent feelings of abandonment and to aid the reconstruction of an individual's life. Reparation for the individual, and consequently confidence in public authorities, also occurs through the engagement of the latter in securing the trial of those responsible in the case of man-made crises (technological disasters and acts of terrorism). Post-crisis communication by public authorities helps maintain cohesion by showing victims that responsibles at State level or local authorities continue to take care of their problems and ensure the social cohesion.

Actions can then be taken in terms of compensation for damage incurred (humanitarian actions, government or international initiatives and insurance companies in the matter of indemnity). Moreover, organizational training, through the detection of failures by experience feedback, can enable lessons to be learned from the crisis. It is then possible to think differently about the reconstruction of affected areas to make them less vulnerable (urban planning for disasters, adaptation of networks, buildings and urban forms against dangers) and crisis management systems can evolve.

The re-establishment phase also calls for a new communication sequence in the course of which contact with the media and partners will be maintained by the crisis team. It enables account to be taken of progress achieved in the implementation of long-term solutions identified during the intervention.

Finally, the need for rapid re-establishment should of course be further anticipated in advance of crises both to limit the duration of the economic, social, psychological or political consequences of a crisis and through reconstructions, to avoid repeating past errors. Considering post-crisis re-establishment through anticipation is, therefore, to envisage an adaptation of socio-regional systems in their entirety, so change can enable them to adjust to future crises, and not just return to the same state of affairs.

8.4. Conclusion

There are many tools for avoiding, mitigating or preparing for crises. Feedback reveals the point at which social organizations (at State, authority and business levels) should have the capacity to anticipate, plan and prepare crisis management. They also underline the importance of communication during and after the crisis, as much in directing emergency services as for the public. Finally, they illustrate the important role of individuals themselves through their capacity for adaptation, reaction and flexibility, whether these individuals are responsible or not.

Recent events show that some crises remain difficult to prevent despite the existence of planning systems, crisis teams or predefined counter-measures depending on the scenarios planned for. From all the evidence, it appears difficult if not impossible to predict the unthinkable, even with the best planning tools and the best simulation models of weaknesses or vulnerabilities. Therefore, inevitable freak events teach us that preparation for crisis is a condition *sine qua non* of resilience. Anticipation enables us to prepare for the worst and not to be paralyzed in the face of complexity, the unknown and uncertainty.

8.5. Bibliography

[ANS 10] ANSELL C., BOIN A., KELLER A., "Managing transboundary crises: identifying the building blocks of an effective response system", *Journal of Contingencies and Crisis Management*, vol. 18, pp. 195–200, 2010.

[BOI 03] BOISVERT P., MOORE R., La gestion des crises et des situations d'urgence: Un guide pour les gestionnaires de la Fonction publique du Canada, Centre canadien de gestion, 2003.

[DOV 92] DOVERS S.R., HANDMER J.W., "Uncertainty, sustainability and change", *Organization & Environment*, vol. 9, no. 4, pp. 482–511, 1992.

[FIN 96] FINK S., *Crisis Management: Planning for the Inevitable*, Backinprint, 1996

[GRA 08] GRALEPOIX M., Le plan communal de sauvegarde. Une approche territoriale de la sécurité civile à travers l'enquête des conditions de mise en place dans les communes françaises, Conseil National de la Protection Civile, Internal report, available at http://halshs.archives-ouvertes.fr/halshs-00406498, 2008.

[KER 95] KERVERN G.-Y., *Éléments fondamentaux des cyndiniques*, Économica, 1995.

[LAG 93] LAGADEC P., *Apprendre à gérer les crises: société vulnérable, individus responsables*, Paris, Éditions d'Organisation, 1993.

[MEN 01] MENONI S., "Chains of damages and failures in a metropolitan environment: some observations on the Kobe earthquake in 1995", *Journal of Hazardous Materials*, vol. 86, pp. 101–119, 2001.

[ROU 00] ROUX-DUFORT C., *Gérer et décider en situation de crise*, Dunod, 2000.

[RUI 04] RUIN I., LUTOFF C., "Vulnérabilité face aux crues rapides et mobilité des populations en temps de crise", *La Houille Blanche*, no. 6, pp. 114–119, 2004.

[STA 03] STACEY R., *Strategic Management and Organization Dynamics. The Challenge of Complexity*, 4th ed., Prentice Hall, 2003.

[UN 05] UNITED NATIONS, National Post-Tsunami Lessons Learned and Best Practice, Workshop, Colombo, Sri Lanka, p. 24, 8–9 June 2005.

[WYB 09] WYBO J.-L., VAN WASSENHOVE W., *Retour d'expérience et maîtrise des risques; principes et méthodes de mise en œuvre*, Tec & Doc Lavoisier, 2009.

9

(Re)Constructing Resilient Districts: Experiences Compared

To better understand the variety of ways to implement resilience, we have chosen to look at New Orleans and East London. These two spaces are the subjects of reorganization-reconstruction projects that hope to construct a "resilient city", able to overcome future environmental disruption, whether that be natural disasters (hurricanes and floods) or as yet unknown consequences of environmental changes in centuries to come (climate changes, a rise in sea level, subsidence, etc.). The two cities can thus be thought of as resilience laboratories.

New Orleans and London both use references to resilience explicitly, but with different meanings. In the case of New Orleans, resilience is understood in a post-disaster context, as a process of *recovery*, which means physical reconstruction as much as the restoration of the social, economic, environmental, political, ideological and symbolic system devastated by Hurricane Katrina. In London's case, the disaster of the Thames flooding has still not occurred: resilient rather means the capacity to absorb future shock. This dimension is not, however, absent in New Orleans, since the reconstruction also aims to create a more secure city based on the lessons learned.

Chapter written by Julie HERNANDEZ and Stéphanie BEUCHER.

9.1. (Re)New Orleans: Big Easy as a resilience laboratory

The disaster that hit New Orleans in the summer of 2005 and the city's reconstruction in the years that followed allow the ambiguities and limits of the concept of resilience to be illustrated, once put into practice. Here, the term is ambivalent from the outset, since it can equally mean the process that leads to the city's reconstruction or restoration after the disaster, and so, by stages, a return to a "normal" state of things, assimilated to a situation similar to what existed before the disaster hit, but also the rebuilding of a city able to overcome future disruption. This ambiguity in the urban project is moreover expressed in the socially and politically charged invocation, "ReNew Orleans", which highlights the delicate semantic balance between the idea of "remaking anew" and that of "remaking from new" from the damp ruins of the city.

9.1.1. *"Rebuild it bigger" and "the optimism of disaster" in the United States*

The question of resilience as a process of reconstruction, or more precisely *recovery*, is asked with regard to the hurricane leading to the radical collapse of the urban system. Katrina's passage in fact affected this system in all its dimensions:

– from a physical point of view, the flooding of more than 80% of the city left significant physical traces in the form of the accelerated decay of buildings and infrastructure (roads and water pipes eaten away by the intrusion of salt water), sometimes replaced by spectacular new constructions (the project Make It Right in the Lower Ninth Ward and the construction of the gigantic LSU Medical Center in the Mid-City quarter);

– from a demographic point of view, the city has undergone a dramatic drop in its population, which dropped from 450,000 inhabitants in July 2005 to around 350,000 in July 2010 while undergoing significant qualitative transformations (a reduction in the average age of the population, an increase in its social, economic and educational level due to the *gentrification* of part of the city, the appearance of a Latin-American minority);

– from an economic point of view, besides the total cost of repairs estimated at more than 70 billion dollars, of which part remains the responsibility of a town council already bled dry before the hurricane's arrival, the consequences in terms of income are still difficult to measure but remain catastrophic in an American context marked by the economic crisis, which has been worsening since 2008;

– finally, the shock left by Katrina on the urban psyche has left permanent trauma, with the memory of the collapse of technical protection systems as well as other safety nets.

Since September 2005, the historic examples of Chicago and San Francisco have been put forward as potential prospects for the reconstruction of New Orleans. The former was devastated by the great fire of 1871, which destroyed a third of the city, including its economic center, and left more than 90,000 people homeless. Less than 10 years later, the city was rebuilt, notably with the first sky-scraper ever built, driven by intense demographic growth. It even became the uncontested economic capital of the American Middle West and the United States' second city at the turn of the 20th Century [MIL 96, HAR 04]. In San Francisco, the great earthquake of 1906 destroyed more than 28,000 buildings, deprived 225,000 people of shelter and caused more than 400 million dollars worth of damage. Yet in 1915, less than 10 years after the earthquake, visitors from the Panama-Pacific International Exhibition noted that practically no sign of the disaster was visible and that the city had modernized its infrastructure considerably (boulevards were widened, tramway lines improved and sewage pipes were rebuilt [BIR 06]).

From this perspective, envisaging "Katrina" as an opportunity to rebuild a demographically larger and economically more powerful New Orleans was a natural reaction, in keeping with the historic notion of "resilience" in the United States.

Except for the fact that comparisons of demographic and economic indicators from the three cities over the course of their respective reconstruction periods ignore the dynamics at work before each disaster. These indicators were only a catalyst of these evolutions. Thus, Chicago in 1871 and San Francisco in 1906 were two cities in the middle of demographic growth, whose competitive industrial sectors and progress in infrastructure (the railway in Illinois and the port in California) attracted skilled labor and investment. In contrast, New Orleans at the beginning of the century had been steadily losing its population for 40 years and its economic capital was concentrated in the area around the port, a poor provider of employment for locals, and the tourism sector was flourishing but vulnerable, as much in its sources of income as in the employment opportunities it offered to the local population.

If the return of inhabitants after August 2005 was presented as the condition *sine qua non* of urban resilience, there was, after the ritual invocation of a "right to return", little discussion about the size of population that the collapsing local economy and infrastructure could effectively support sustainably, and still less on the exact number and spatial distribution of people that technical protection systems could effectively protect from a future disaster.

9.1.2. *"Rebuild it safer": impossible strategies for urban retreat*

The most standard approach defines a social system's resilience capacity as "the system's capacity to organize itself to be in a position to draw lessons from past disasters to protect itself from them better". The city's reconstruction should, therefore, create a more secure urban system if needed.

In 2008, the approach of the less powerful Hurricane Gustav suggested that technical reinforcement measures for protection systems had at least rebuilt the latter to their pre-2005 level. The closure of the Mississippi River Gulf Outlet and the installation of flood gates at the entrances to evacuation channels for water pumped into Lake Pontchartrain, (previously perfect Trojan horses channeling storm surges into the heart of the city), represent unmistakable progress in the immediate defense systems against future storm devastation. But "*this short-term mania for landscape engineering solutions*" [MAN 08] rests on a technical martingale producing ever higher, wider (and more expensive!) levees to compensate for the inexorable subsidence of the city and the disappearance of marshy areas in front of the delta. This approach, which makes the urban defense lines more rigid, contradicts an acceptance of the latter as a system of environment–human relations that should be sustained flexibly to ensure its adaptive capacity is more sustainable across the transformations of different environments [MAR 08].

This notion of flexible urban planning has often been put forward in debates informing New Orleans' resilience after Katrina. Some have adopted the perspective of the urban material and "floating city" projects have multiplied, as architecturally attractive as they are preposterous for the reality of urban buildings more than 300 years old. But the literature focuses more often on urban form and the need to leave "breathing spaces", incorporated in case of flooding through the demarcation of "no-building" zones and the creation of spaces for overflow and excess water collection.

The hypothesis of rebuilding a "more secure" city, not through competition in technical investments but through reorganization, often synonymous with a strategic reduction of the urban footprint, fits perfectly into current reflections on the sustainability of cities. Whether it is a question of limiting cities' exposure to external hazards or streamlining the relationship between the environment and their own development, the hypotheses put forward run counter to the progressive narrative that measures resilience only in terms of urban growth.

However, the debates and challenges governing the reconstruction of the quarters of New-Orleans have in turn constituted a sharp empirical challenge to this rationalistic assessment of the notion of resilience. If sustainable improvement of the city's resistance capacity to future disasters seems effectively subordinated to the strategic reduction of the urban imprint and the concentration of populations and resources, the "big footprint debates" [CAM 08], which ran through all of New Orleans society between 2005 and 2007, has rapidly highlighted this proposition's incompatibility with an interpretation of urban resilience that suggests a "better" city would necessarily be "fairer".

9.1.3. *"Rebuild it fairer": how to negotiate "resilience" and "spatial justice"*

Socio-spatial and environmental inequalities revealed by Katrina have intensely energized the discussion underway on the ideal of "urban justice" as the ultimate objective of all planning processes [FAI 09]. Although the socioeconomic and racial distinctions between the spaces and populations affected by the disaster were not as clear-cut as media accounts have implied [HER 10], it still remains that New Orleans' urban system was made vulnerable, above and beyond its physical relationship to its environment, by a series of social and economic problems (astonishing crime rates, the mediocre quality of the public education system, large income gaps and the systemic perpetuation of conditions of extreme poverty for some of the New Orleans underclass [GER 07]). In the face of this, it seemed obvious that any urban resilience measures after Katrina should involve some indication of a lessening of its entropic social character, which created conditions of acute marginalization, exposed when the hurricane passed.

In the context of the post-disaster situation, "*the reconstruction of affected cities proposes not only to remedy the loss of material goods, but also the inequalities and injustices perceived in these losses*" [STE 08]. Consequently, New Orleans' resilience could not be limited simply to replacing damaged goods and resources but should necessarily include mitigating urban populations and districts' differential vulnerabilities. The implementation of this "*necessary attention to the needs of the most vulnerable populations, which we call 'reconstruction' is fundamentally guided by ethical questions of equality*" [VAL 05] has been envisaged in two ways. In one instance, by measuring the concrete investments made in terms of reconstructing essential urban resources and services in the most deprived districts: the number of schools, medical centers, trade centers and social spaces reopened after Katrina would suggest a return of "the city" to previously marginalized spaces. At least, the reconstruction of public transport and means of access to the city signal these districts' integration into the urban fabric. A second possibility involved bringing back the most vulnerable sectors of the city, through a voluntary socio-spatial mixing policy in terms of housing, jobs, services and leisure. The diversity promoted in these districts would in this case allow the promotion of "numerous and quality interactions" necessary to maintaining and adapting the urban system in its entirety [ASC 00].

These two perspectives (the redistribution of resources or the redistribution of people) are not necessarily incompatible with resilience logistics via "urban growth" or via the "spatio-environmental rationalization" described above. However, they pose a more subtle problem in evaluating the measure in which "spatial injustice" is not only a series of urban malfunctions that should be resolved, but also a feeling shared by those who suffer by it. Hence, it is necessary to mend not only *"the tears in the urban fabric"* [CAM 08], but also the urban identity and legitimacy of its creators.

The interpretative lines for the notion of resilience, promoting a bigger, safer or fairer city, are thus not necessarily compatible: neither the environment nor the local economy allowing anyone to envisage rebuilding the city larger than before the disaster. The reduction of urban vulnerabilities even supposes a rational decline in the urban imprint. But the spatial justice perspective, embodied in the rarely disputed affirmations of a "right to return", is at odds with all interventions claiming to improve urban resilience by redistributing populations and resources away from the most vulnerable districts.

In many ways, the geographies of New Orleans after Katrina are the product of this dilemma, in which resilience on one level is interpreted as an injustice on another.

9.1.4. *Resilience is not always a good thing*

The reconstruction of cities has motives that urban planning agencies' rationale sometimes seems to ignore. New Orleans has thus been envisaged as an immense urban planning and redevelopment laboratory, both "resilient" and "sustainable", a rich scientific production center sometimes doubled by urban experiments based on the hypothesis of a *tabula rasa* offered by Katrina.

Yet, several years after the disaster, the city's reconstruction has been the product only of very ordinary forces, organized by the most powerful one of all, perhaps: the unchanging nature of the production and reproduction dynamics in the urban space. These have in fact backed up the almost instinctive temptation to rebuild the city "from memory" which has been seen before, in the case of other cities devastated by natural disasters [KOF 05].

The lack of change in the structures and their identical or near identical reproduction can be seen as an elementary form of this resilience. Identical reconstruction appears to be the least expensive process on all levels of urban planning. In economic terms, it circumvents the expense of compensating displaced people and reorganizing activity. At the level of urban communities, it rests on the hypothesis of a complete recreation of social networks that will come to support resilience in the long-term. From a political point of view, it removes the need to make drastic planning decisions. Finally, from a cultural point of view, the identical reconstruction of the ruined city arises from the temptation to overcome trauma purely and simply by effacing its traces in the landscape.

This inclination toward *laisser-faire*, driven by an instinctive desire to return to the normal state of affairs before the disaster, is usually accompanied by the construction of a discourse on a "Golden Age" before the disaster and the development of a selective memory of the disaster and the conditions that led to it [HER 08]. This memory was all the quicker to dominate the reconstruction process in New Orleans, as the city's survival relied, even before the disaster, on discourses of inertia, especially because the city's tourist industry depended almost entirely on a picturesque reinvestment in history and the urban cultures of the past [STA 06].

Whatever the indicators used to measure it, the impression of a "piecemeal" reconstruction dominates observations of the city. The rhythm and scale of this reconstruction have been very different and often incoherent from one district to another, but also from one street, or even one house to another, multiplying the problems resulting from the decay *in situ* of damaged buildings that put a strain on districts' re-establishment. With a population stabilized at a little under 80% of what it was before the hurricane (343,829 inhabitants at the 2010 census) on a similar urban footprint, the city gives the impression of floating in a distended urban fabric for which it is increasingly difficult to ensure an effective and efficient distribution of public services. The gaps in water supply and sewage services, the disappearance of some bus services that ensured connections into the most marginalized neighborhoods and the absence of a public hospital after the controversial closure of the Charity Hospital are so many factors that contribute to re-victimizing the weakest populations. In addition, the accumulation of crises resulting from the blight of decaying buildings, from an already poorly performing public education system, in particular for the city's most deprived students, and rampant crime are pushing New Orleans further and further toward the model of cities in the North and East of the country, marked by the slow and silent disaster of deindustrialization and economic decline.

In this sense, Katrina is not likely to have produced a bifurcation. It was rather instrumental as a catalyst in the urban trajectory of New Orleans, the first American city to be struck by a disaster of this size where all economic and social indicators were already flashing red; and the most meaningful "resilience", if by that we understand the survival of certain dynamics despite the occurrence of disasters, was paradoxically that of its dysfunctions. To the extent that, far from the attractive analogy with Port-au-Prince, which would allow the urban tragedy revealed by Katrina to be portrayed as an exception in a North-American context, New Orleans' path was catalyzed by the disaster and directed toward the urban models and issues that Detroit, Cleveland or even Buffalo have been facing for more than two decades.

9.2. Urban renewal and resilience in East London: the Thames Gateway

A very different case *a priori*, the Thames Gateway is a vast and ambitious urban renewal project, which extends for 70 km downstream of the Thames, from the borough of Tower Hamlets to its estuary. This area has undergone significant reincarnations. It was home to most of the infrastructure for the 2012 Olympic Games. More broadly, predicted construction in the Thames Gateway area of London for the next 10 years represents more than a quarter of Greater London's housing needs.

The planning documents adopt a sustainable development perspective: they promote a compact city, with dense housing, with mixed populations and functions. Beyond the classic challenges of urban renewal, the districts concerned are subject to flood risk both from the river and the sea [BEU 08, MCF 09]. These floods are rare since these districts are well protected by flood defences. However, their survival is problematic and explains the implementation of resilience strategies.

As with New Orleans, resilience as a process is glossed over in favor of resilience as a property, which additionally expresses the transition from the substantive (urban resilience) to the adjective (a resilient city or district). This is not about evaluating the forms and consequences of a reconstruction process, although it would be good in the *tabula rasa* situation evoked for New Orleans, but of understanding resilience as the ability to absorb future disruption.

9.2.1. *A development project: a beacon in the context of major risk*

The Thames Gateway aims for the "rebirth" of entire sections of urban London in districts characterized by deindustrialization, insecurity and unemployment, by transforming 38,000 hectares of brownfield into new pillars of development. The Thames Gateway is becoming a laboratory for putting theories on urban renaissance into practice. Compared with earlier experiments in urban regeneration carried out in Great Britain, the novelty lies in the fact that the scale of reflection and action is no longer a limited area. Urban renewal invites us to consider the city on many levels (on the level of the city, district, project, etc.) and involves coordinating many municipal services.

However, paradoxically, the definition of integrated urban projects does not induce stakeholders to take account of floods. Nonetheless, the risk is not negligible: if the flood defences were to be broken or if a wave were to pass over the protection infrastructures, the consequences would be dramatic, with more than 6,200 hectares situated below sea level, in Thamesmead, for example. This would result in hundreds of deaths. In 1928, 14 people were drowned in this way in Westminster and the 1953 storm took more than 300 lives in East London, with waves nearly 3 m high at the entrance to the estuary. The flooding then covered 11,600 hectares within the city and affected 20 boroughs.

Additionally, although the risk is slight, the impacts of flooding would be considerable, as the stakes are significant and likely to increase with urban renewal projects. According to the Environment Agency, 500,000 properties, 400 schools, 16 hospitals and numerous underground stations, fire stations, sewage treatment plants, generating stations and gas production plants as well as ports are exposed to flood risk in the Thames Gateway. Before flood defenses were constructed and before the Thames Gateway project, it was estimated that 15% of London's population were exposed to centennial floods. However, the size of the stakes is quite different if new development projects are taken into account. According to the Association of British Insurers (ABI), 91% of projected new housing is situated in zones at risk of flooding [ASS 05]. A major flood in the Thames Gateway could cost between 12 and 16 billion pounds, with the sum increasing by 4 or 5 billion if new developments are taken into account. The ABI evaluates the value of assets lying in the capital's flood zone at 126 billion pounds and predicts that this figure will increase by 19.1 billion pounds with new developments [ASS 05]. It has also estimated that providing insurance cover for new housing situated behind the levees would cost insurers 26 million pounds, but that could reach 195 million pounds if the sea level rises by 40 cm.

9.2.2. *From sustainable development to resilience*

Initially, the issue of floods was paradoxically not integrated into the urban renewal project. Flood management policies are defined on a national level, not at the level of the London metropolis. Although, according to PPS25, local authorities are supposed to integrate the risk into their urban development plans, carry out Strategic Flood Risk Assessments across their districts and refuse construction permits in zones prone to flooding, in practice their flood management proceedings have been shown to be generally uneven and weak.

This failure to take risk into account in urban projects can be explained as much by the lack of training for developers as by the feeling of security that dominates in the whole urban region due to the existence of a network of complex flood protection infrastructures: levees put in place along the river bank as well as the large mobile barrage constructed downstream from the capital, the Thames Barrier, protect the population from this risk. Other mobile barriers, such as the Barking Barrier at the mouth of the River Roding or those at the entrance to the basins of the Royal Docks, have also been implemented in the second half of the 20th Century. According to the Environment Agency, the levees guarantee a degree of protection against a flood with a 0.5% chance of occurring, and the Thames Barrier will protect the capital against a flood with a 0.05% chance of occurring, until 2030. Second, according to current estimations of the rise in sea level, the mobile barrage will protect Londoners against a millennial flood [DEF 05]. In the minds of a large number of stakeholders, including those behind the Olympic project, everything is happening as if constructing new districts along the Thames front or its tributaries and the arrival of populations attracted by the new environmental amenities on the redeveloped river banks will naturally define districts capable of adapting to floods. For a large number of developers, renewing districts on the river banks allows the link between the inhabitants and their river to be recreated and makes life beside it possible in all circumstances. They also think that the associated environmental measures will suffice to prevent or at least limit the risk.

Later, obtaining the July 2005 Olympic Games lent new scope to the Thames Gateway project. The flood risk has, therefore, been evaluated (in the Flood Risk Assessment published in January 2007 and revised in May 2007) and the new development's vulnerability discussed. The desire to protect the Olympic Park can be explained by the significant media attention given to the consequences of Hurricane Katrina. Moreover, adapting to potential crises became a pressing need after the attacks on 7th July 2005. The need to implement all hazard resilience processes became a major challenge for the Greater London authorities.

Moreover, the floods that affected a large part of the country in July 2007 have contributed to raising awareness of the capital's vulnerability in the face of large-scale disaster. If London was relatively little affected by the event – although localized flooding from urban run-off was able to cause significant damage – the expert report ordered by the government [PIT 07] is significant: the flood management policy had a number of weaknesses at all levels, especially at crisis management level. The report emphasized the Thames estuary's growing vulnerability, as the Thames Barrier barrage is becoming less effective.

Resilience is thus the obvious solution. In the field of flood prevention, most stakeholders are now aiming for resilient constructions (i.e. constructions that allow inhabitants to return less than 3 weeks after a disaster) and not just resistant ones (using impermeable materials that prevent water entering, but which can worsen flooding around the building). The term "resilience" is also increasingly used by crisis managers, for example, when applying the 2004 Civil Contingencies Act, in a fairly precise sense: it is the capacity to recover from major crises: attacks, epidemics, floods, etc. The objective is not merely to reinforce the country's capacity to manage crises, but also to predict and anticipate them through work begun well before the crisis.

9.2.3. *Resilience as a multiplicity of crisis preparation structures*

In the run-up to the Olympic Games, preparation structures for a major crisis in the London metropolis multiplied. The most significant of them is the London Resilience Partnership, an organization that ensures coordination between the government, the Mayor of London and all the capital's crisis management bodies (local authorities, economic areas, the transport sector, the voluntary sector, etc.). Its strategic programs are defined by the London Regional Resilience Forum (LRRF) which detailed the London Strategic Flood Framework in January 2010. This seems to gather together a large number of decision makers and to contribute to an improved understanding of floods. In the July 2011 London Plan, there also figures the rubric "Safety, Security and Resilience to emergency".

The London Resilience Team represents the LRRF's operational component, aiming in the case of crisis to ensure services' rate and effectiveness as well as to minimize the consequences for the entire country. Its president directs the London Regional Resilience Program Board, which develops regional priorities and ensures that resources are sufficient to reach the defined objectives. The London Risk Advisory Group prepares the regional risk assessment and the community risk registers to be sent to the six Local Resilience Forums which should list all types of information on all the risks that might affect the London metropolis and create databases useful to developers. The Local Resilience Forums constitute information collecting bodies rather than enforcement structures. Today, however, these forums seem only to exist on paper: in October 2010, the London Regional Resilience Forum became the statutory body for taking action in the field of resilience, whereas the Local Resilience Forums are no longer institutionally recognized.

9.2.4. *TE2100: a resilience strategy*

At the same time as improving crisis management structures aiming to construct a resilient city, the Environment Agency has raised concerns about the potential consequences of the Thames Gateway project on the increasing vulnerability of the Thames Valley and since 2006 has put in place a new project: TE2100.

The goal is to detail a management tool and to aid decision-making in order to reduce the vulnerability of London and the whole of the Thames estuary facing a flood risk likely to be increasingly worrying if predictions on climate warming and the potential rise in sea level prove correct. The Environment Agency's area of study is the Thames flood plain where the impact of the tide can be seen, that is to say the whole estuary up to Teddington, west of London.

TE2100 is a risk adaptation plan more than a risk reduction plan. The studies rely on the results of the project Foresight, a study of climate warning's impacts on the United Kingdom. It is especially concerned with implementing integrated management of major flood plains in order to find a compromise between their urban development and taking account of the flood risk, mainly by increasing water storage capacities in areas of marsh and plains at low altitude. The idea is to develop spaces that can be flooded in case of extreme events with a low probability (1%) of occurring in order to reduce the water level of the Thames by 80 cm. This option would enable the functional life of the Thames Barrier to be increased by 50–70 years. Moreover, if buildings that are resilient for 20 or 30 years are designed with the potential to be flooded later, this will involve offering citizens the chance to acquire land for 25 or 30 years. These suggestions are hardly realistic but they show the need to reflect on the resilience of urban projects. TE2100 finally involves a partnership with insurance companies. How do we insure a property which we know may be voluntarily flooded in case of an extreme event?

TE2100, therefore, invites us to reconsider the development and planning system in its entirety. It calls definitively for a radical modification in culture. Although the term "resilience" is still little used (or used in the weak sense of using materials allowing a rapid recovery from floods), the project constitutes a real reflection on resilience. It is no longer a question of trying to prevent or reduce risk but of learning to live with it. The objective is not to fight risk but to consider it as a structural component of some districts and to find solutions to adapt to it at all levels. With the TE2100 project, the Environment Agency proposes somehow to create a risk district that draws its identity from its resilience, that is to say from its capacity to live with the river. The goal is to overcome divisions between urban development and hazard mitigation, between prevention and crisis management, and between national and local action to make society at large understand that all sectors at all levels can improve, at their own level, the resilience of goods and people, that is to say the capacity to live with floods. This involves thoroughly informing populations and bringing them up-to-date so that they are in a position to decide the degree of risk acceptable to them.

9.3. Conclusion

These two examples allow resilience to be put to test on the ground. Whether resilience is understood to be a post-crisis reconstruction process or an intrinsic property becoming the horizon for development projects, it is clear that its implementation is not to be taken for granted.

Speaking of reconstruction, the debate quickly shifts into measuring the post-disaster evolutions, the difficult balance between identical reproduction, which perpetuates earlier urban vulnerabilities, urban dysfunctions, injustices and change, which might lead the urban trajectory to bifurcation. The case of New Orleans shows in particular the importance of heritage and the weight of the dynamics that shape the urban fabric before the crisis, making it more a catalyst than an actual driver of change.

Speaking of resilience as a strategy, a project aiming to develop a social and regional property, its implementation depends on all the environmental, socioeconomic, cultural, political and institutional characteristics of the region.

In both cases, the question of scale is instrumental, whether these are temporal levels, with the question of differential temporalities and rhythms for reconstruction and renewal, the weight of heritage, the differing impacts over time or spatial scales, with the search for perimeters for reflection and actions or the multi-scalar and trans-scalar dimension of the questions considered. The (geo)political and organizational dimension is equally critical: these two examples show the importance of the role of developers, government power, private partners, individuals, communities, etc.

9.4. Bibliography

[ASC 00] ASCHAN-LEYGONIE C., "Vers une analyse de la résilience des systèmes spatiaux", *L'Espace Géographique*, no. 1, pp. 67–77, 2000.

[ASS 05] ASSOCIATION OF BRITISH INSURERS, Making Communities Sustainable, Managing Flood Risks in the Government's Growth Areas, Report to Association of British Insurers, Entec, London, February 2005.

[BEU 08] BEUCHER S., Risque d'inondation et dynamiques territoriales des espaces de renouvellement urbain: les cas de Seine-Amont et de l'Est londonien, 2008.

[BIR 06] BIRCH E.L., "Learning from past disaster", *Rebuilding Urban Places after Disaster. Lessons from Hurricane Katrina, Philadelphia*, The University of Pennsylvania Press, pp. 132–148, 2006.

[CAM 08] CAMPANELLA R., *Bienville's Dilemma. A Historical Geography of New Orleans*, University of Louisiana Press, Lafayette, LA, pp. 344–350, 2008.

[DEF 05] DEFRA, Making Space for Water: Taking Forward a New Government Strategy for Flood and Coastal Erosion Risk Management in England: Government First Response, Department for Environment, Food and Rural Affairs, London, 2005.

[FAI 09] FAINSTEIN S.S., "Spatial justice and planning", *Justice Spatiale/Spatial Justice*, pp. 58–77, vol. 1/2009 available at http://www.jssj.org/wp-content/uploads/2012/12/JSSJ1-5en1.pdf.

[GER 07] GERMANY K.B., New *Orleans after the Promises: Poverty, Citizenship and the Search for the Great Society*, University of Georgia Press, Athens, 2007.

[HAR 04] HARTER H., "Chicago et l'incendie de 1871: entre mythes et réalité", in CABANTOUS A. (ed.), *Mythologies urbaines. Les villes entre histoire et imaginaire*, Presses Universitaires de Rennes, Rennes, 2004.

[HER 08] HERNANDEZ J., "Le tourisme macabre à La Nouvelle-Orléans après Katrina: Résilience et Mémorialisation des espaces affectés par des catastrophes majeures", *Norois*, vol. 208, 2008.

[HER 10] HERNANDEZ J., ReNew Orleans? Résilience urbaine, mobilisation civique et création d'un "capital de reconstruction" à la Nouvelle Orléans après Katrina, Thesis, University Paris Ouest Nanterre, 2010.

[KOF 05] KOFMAN B.C., ULLBERG S., HART P., "The long shadow of disaster: memory and politics in Holland and Sweden", *International Journal of Mass Emergencies and Disasters*, vol. 23, pp. 5–26, 2005.

[MAN 08] MANAUGH G., TWILLEY N., "On flexible urbanism", in STEINBERG P., SHIELDS R. (eds), *What is a City? Rethinking the Urban after Hurricane Katrina*, The University of Georgia Press, Athens, pp. 63–77, p. 67, 2008.

[MAR 08] MARET I., GOEURY R., "La Nouvelle-Orléans et l'eau: un urbanisme à haut risque", *Environnement Urbain/Urban Environment*, vol. 2, pp. 107–122, 2008.

[MCF 09] MCFADDEN L., PENNING-ROWSELL E., TAPSELL, S., "Strategic coastal flood-risk management in practice: actors' perspectives on the integration of flood risk management in London and the Thames Estuary", *Ocean & Coastal Management*, vol. 52, pp. 636–645, 2009.

[MIL 96] MILLER D.L., *The City of the Century: the Epic of Chicago and the Making of America*, Simon and Schuster, New York, 1996.

[PIT 07] PITT S.M., The Pitt Review, Independent Review Issues Call to Action on Flood Risk, 17 December 2007.

[STA 06] STANONIS A.J., *Creating the Big Easy: New Orleans and the Emergence of Modern Tourism*, The University of Georgia Press, Athens, 2006.

[STE 08] STEINBERG P., SHIELDS R. (eds), "Memories", *What is a City? Rethinking the Urban after Hurricane Katrina*, The University of Georgia Press, Athens, pp. 125–128, 2008.

[VAL 05] VALE L.J., CAMPANELLA T.J., *The Resilient City. How Modern Cities Recover from Disaster*, OUP USA, 2005.

10

Resilience, Memory and Practices

Resilience, seen as a discourse and a political construct, refers equally to the question of how risk is perceived and remembered, and above all to post-disaster memorial messages. Indeed, perceptions and memories[1] of disaster hold a crucial place in the complex network of processes and actions that the concept of resilience encompasses. Risk management, which is the main objective of resilience[2], is largely based on the perception of risk and the memory of disasters. So what policies and developments can facilitate resilience in terms of the memory process? Which memory should be emphasized to initiate or deepen urban resilience strategies?

These questions might seem rather simple – it could be a question of making the traces of disasters part of local heritage (patrimonializing them)[3] in order to cultivate their memory and facilitate management of future occurrences. However, there are innumerable examples of failed risk management despite the persistence of traces, warnings, memorial policies and risk awareness – for example, among other events, the destruction of the city of L'Aquila in Italy following the April 2009 earthquake,

Chapter written by Antoine LE BLANC.

1 In this chapter, we will use the terms "memory" and "perception" in their broadest and most common sense. However, for conceptual clarification, refer to [HAL 50] and [BEC 92].

2 The concept is defined accordingly in [COM 10].

3 To better understand the notion of heritage and its dynamics, see [CHO 92].

although the region's history of earthquakes was prolific and well documented.

10.1. The resilient system between identity and evolution

The varied definitions of "resilience" emphasize the system's capacity to return to what it was and to persist in its identity despite major disruption. Yet, by definition, its path has been disturbed by the disruption, and therefore the system's identity is not exactly the same after the disaster.

Such is the resilient system's paradox: it should simultaneously absorb the shock to remain the same, and take account of it to evolve and henceforth learn how to better resist this type of shock. The whole challenge lies in the gap between the system's flexibility and its identity. From what moment and according to what criteria can it be decided that the system's qualitative structure has been modified and that this system is, therefore, not resilient?

We have no definitive answer, but we suggest putting it in a slightly different way. Resilience, as we understand it, is above all based on the local, associated places, people and dynamics. We focus on the system's variability: it is not about making a static snapshot of a district, but of complex, qualitative and evolutionary dynamics. It is necessary to understand which aspects of the system have been destroyed by the disaster and which structures can be reconstructed by focusing on all stakeholders and their memory: isn't it resilience essentially when individuals survive and communicate their memories? Does that suffice to conserve a system's qualitative structure?

10.1.1. *Resilience and district identities*

This multiple memory, based on a wide range of individual perceptions, is difficult to transmit. It is precisely this complexity that enables the urban system to be resilient. Indeed, risk management carried out from a resilience perspective relies on a nuanced, complex, delicate and qualitative perception of risk, the perception resulting from the involvement of a number of important stakeholders, on various scales.

The construction of a shared memory, the entangling of these perceptions, in the quality of relationships between stakeholders: the way in which they communicate and share an understanding of risk, is instrumental to resilience. The actions undertaken will depend on the information shared, itself the result of varied perceptions of risk. This is why Rhinard and Sundelius [RHI 10] have been able to show that resilience depends on the quality of coordination between stakeholders, arising from the means of cooperation and how well acquainted those players are.

In practice, the goal of resilience will lead to less being spent on major technical and structural approaches, and more being spent on educating the public and installing dynamics for knowledge-sharing, confidence and cooperation. Resilience thus involves partially delegating risk management to stakeholders other than the conventional political decision-makers. *Top-down* structures are not replaced but weakened, completed by *bottom-up* structures, which end by democratizing the process of informing and deciding, and enable the complexity and variety of perceptions of risks in the population to be taken into account. We can think, for example, of the reconstruction of Friuli after the devastating earthquake of 1976: a reconstruction considered to be exceptionally rapid and effective, although an earthquake had not been expected in that region. The in-depth study of reconstruction dynamics has shown that they were largely initiated by the inhabitants without waiting for the authorities to react, and that resilience – although the word was not used at that time – was stronger where solidarity between inhabitants, the relationship dynamics and confidence and the quality of the links between districts functioned to the full [FAB 86].

Resilience has thus been strongly linked to regional identity, to local knowledge that is habitual, lived, felt and known; resilience increases proportionally to better knowledge of the region and its inhabitants, to the quality of the relationships and interactions between stakeholders.

10.1.2. *Temporalities and memories*

Shared memory constitutes the basis of a district's identity, and resilience is based on territory and memory. The memory of place, the functioning of a regional system and regional habits, contributes to consolidating places' and inhabitants' resilience.

This memory is also the memory of disasters. It enables the system to be continuous, depending on various temporalities: the temporality of the crisis, the temporality of risk management, the temporality of human lifespans and inhabitants' generations, etc. Often, however, for various reasons (cognitive dissonance, trauma, forgetfulness), the memory of crises is brief and it differs considerably [TIS 07] between individuals [BAI 96]. Managers are, therefore, confronted with the question of how to perpetuate this memory: how should it be sustained, how should it be kept alive? One of the most up-to-date and effective responses is to give a material, spatial existence to the memory by absorbing the catastrophe into local heritage.

Individual memories are thus projected onto physical, material, constructed symbols. These constitute a visible support, small or large, of the district's memory and identity. Berlin has kept its memorial church (*Gedächtniskirche*) in a state of ruin, Paris has preserved the integrity of its bullet-riddled walls, Japan places markers indicating the reach of tsunamis, and numerous cities have street signs indicating the floodwaters' highest points, etc.

However, individual memories remain vital: material support for memory does not alone suffice to convey a message of the memory of the crisis and risk prevention. Georges Prévélakis has shown an instance in which resilience was stronger after the destruction of a building that occasioned a significant number of victims than after a catastrophe that caused numerous deaths, without damaging any buildings [PRE 10]. A local system recovers more quickly after an event that causes destruction of buildings but few deaths than after a disaster that causes significant mortality. We can refer to regions habituated to frequent but not very violent eruptions, such as La Réunion and Sicily, or the flooding around the Aude or the Somme in France, and compare them to the 2009 earthquake in L'Aquila, which was not the most violent that that region of central Italy had undergone historically, but which had 300 victims. It is also worth mentioning the instances of streets, districts or cities which, despite a change of name or appearance, retain their former name, their former structure and, indeed, their former purpose in the minds of their inhabitants. Istanbul is still both Byzantium and Constantinople. Numerous cities – or countries – are still known by names that are no longer official following regime change or catastrophe; French villages and cities are littered with roads whose names on the sign posts do not correspond to the current, traditional name. Resilience thus seems to rely more on the memory of individuals than on the permanence of places, even if the places, monuments and regional systems constitute powerful supports for identity and memory processes.

10.1.3. *The conservation of ruins: an example*

Analysis of the conservation of traumatic ruins shows that this choice is a resilience tool [LEB 10]: the urban system integrates trauma rather than effacing it, and this choice aims overtly to promote risk reduction, as in the case of Gemona, a small town in Friuli that chose to conserve the ruins caused by the 1976 earthquake unaltered to keep the memory of the disaster alive and develop a risk culture. Here, therefore, the ruin's conservation is a strategy developed from "proactive" resilience, in the sense that the objective is not only to look back, but also to implement steps for prevention with the goal of reducing future risk, in opposition to "reactive" resilience [DOV 96]. The American geographer J. B. Jackson [JAC 05] underlines this message from the ruin using imagery, by comparing the ruins with reminders for unpaid bills from telephone companies: "*de bonne ou de mauvaise grâce, nous attrapons notre carnet de chèques pour acquitter notre dû et éviter ainsi de nouveaux désagréments*" (with good or bad grace, we reach for our check book to do our duty and so avoid new disagreements). Traumatic ruins would be a kind of urban post-it note, an instrument for memory and resilience.

However, conserving a ruin involves the memory of the event, not of the risk; it is, therefore, only *one* element constituting a risk culture.

Indeed, being aware of a past disaster and imagining that that disaster could happen again are two distinct mental processes. It is not enough to secure and conserve a ruin to ensure a proactive resilience process. The ruin should also be presented, placed in context and interpreted [LAC 07]; the ruin's audience should be brought to understand that the risk is still current, using panels or other teaching tools.

It is necessary to develop the space around the ruin. A ruin is a sort of scar: it is the redevelopment of the surrounding space that aids the understanding that the trauma is integrated and overcome, and not simply undergone. In Gemona as in London (the churches of Christchurch, Greyfriars and St Duncan in the East, destroyed by bombing in World War II), local councils have transformed the ruins, while still conserving them, into urban parks and memorials. At Palermo, a ruin in the center of the city (the church of the Holy Spirit) became a concert venue. In Berlin, the Memorial Church has regained its original function due to an architectural construction that complements the ruin. Hamburg developed a small war museum in the monumental ruins of the church of Saint Nicolas.

At Gemona, 30 years after the earthquake, an enquiry we carried out among the city's inhabitants shows that the choice to conserve the traumatic ruin in order to remember and to create a risk culture has borne fruit. Only 11% of the city's inhabitants think that no visible trace of the earthquake remains, whereas 42% of them remember the church of Saint Mary of the Angels, which is, however, not particularly central or important among the city's churches [LEB 10]. The choice of audacious urban planning carried out by the town council really seems to have contributed to the local population's risk culture, and to have accelerated a process, i.e. a proactive resilience.

10.2. Resilience and retaining a memory of risk

Absorbing traces of disasters into local heritage is a complex dynamic, political in nature, which does not automatically succeed in reinforcing urban systems' resilience.

10.2.1. *The ambiguities of preserving traces of catastrophes as heritage*

Since the first tombs were built, mankind has constructed monuments or preserved traces of particularly tragic and sad events [HAL 50], from memorials recording the names of victims of war to buildings conserved as dramatic evidence of events, such as slave houses in colonial cities, or the concentration camps of the Nazi regime.

In almost all cases, this absorption of the disaster into local heritage is accompanied by an ideological and moral message, an injunction to support peace and not repeat the disaster [JEU 01]. This conservation of the memory of trauma, conveyed spatially and elaborated by presentations and interpretations, in theory enables an awareness of risk to be maintained and tends to reduce the exposure to risk in the systems in question.

However, making the disaster part of local heritage is very complex and the political message should be understood in depth, as the example of the concentration camps, as analyzed by the historian Alain Sinou, shows [SIN 10]. Thus, since 1946 the Polish state has made the site of Auschwitz into a museum. Placed on the United Nations Educational, Scientific and Cultural Organization (UNESCO) World Heritage List in 1979, the site today welcomes 1.3 million visitors each year. But why make Auschwitz part of our heritage, and not the other camps? Admittedly, it is the biggest camp and as such it is not just imposing but, equally, representative of horror on a massive scale; but it also supports a political discourse. In fact, until the 1980s, the camp was presented as a place symbolizing the extermination of communists, and not Jews.

Numerous other examples illustrate the politically directed exploitation of traces of disasters. At Reims, after the destruction of World War II, a controversy broke out concerning the cathedral and its possible conservation as it was, gutted by the bombardments; the argument in favor of conservation is not a "never again" desire for peace, but the wish to stoke the fire of hatred against the Germans [AND 86]. More recently, the creation of an Islamic culture center on the ruins of the World Trade Center in New York has also given rise to enormous controversy. The message issuing from the retention of memories of disaster in local culture is sensitive and requires careful handling. In this sense, resilience is clearly a political construction.

10.2.2. *The symbolic construction of disaster and risk*

If resilience is a political construction, it is because disaster and risk are themselves symbolically and politically constructed. Thus, depending on particular cases, it is clear that an event can be constructed to a greater or lesser extent as a disaster, disproportionate to other dramas that are more or less similar, but lived differently, or in different eras. It is particularly instructive to compare media reports of earthquakes and their consequences in different countries, such as Algeria, Italy and Japan. The earthquake in L'Aquila, already described above, had several hundred victims and was the object of intense international media coverage. Yet, its intensity was relatively limited in comparison to earlier earthquakes in the region, indeed compared to other earthquakes experienced in Italy in the last three decades. It is not a question here of denying the drama, but of stating that it is constructed socially and symbolically in a nonlinear manner relative to objective data on the catastrophe.

In the case of L'Aquila, as in many other examples, it is the collective memory relayed by the media that constructed a disaster out of proportion to the measured reality. This discourse amplifying the hazard and, therefore, the disaster leads to the consolidation of a specific memory, to an identity torn apart by the disaster, despite the system reconstructing itself into a state of affairs very similar to what it was before the disaster.

The political and symbolic construction of disaster is a basic element of resilience, or its absence. Here, we could refer to the multitude of examples of human and urban dramas about which information is not repeated and which are not constructed as disasters, which do not create a risk culture and which lead to the same situations being repeated.

10.2.3. *Resilience and forgetting*

Analysis of the link between memory and resilience leads to questions about the right to forget and the reasons why the memory of disasters is short. Individuals, once they have overcome the disaster, often wish to move on to something else. This is the famous paradox for the traumatized: is it worse to remember, or to forget?

We find here the importance of physical and material support for memory. Material support means that the individual is relieved of the burden of remembering. Plato said to Socrates that writing was at once a remedy and a poison because it constituted an external support for memory: remedy, because the memory made external is conserved; poison, because the existence of external support invites laziness in the individual memory. Like writing for Socrates, the materialization and patrimonialization of traces of a disaster enable individuals to remember without having to remember permanently. Resilience is thus constructed on a compromise between memory and psychology, in which inhabitants can remember, due to a monument that revives their memories, but also have the right to live without being constantly confronted with the memory of the trauma.

10.3. The problem of identity

Resilience rests, to varying degrees, on material and immaterial supports (buildings, or memory, message and risk awareness). The political character of resilience leads to questions about risk, the ideological instrumentalization of identity and urban memory, under the cover of a resilience process with positive connotations.

10.3.1. *Resilience confronted by the urban palimpsest*

During post-disaster reconstruction, urban players are confronted with the "palimpsest" dilemma: places are rebuilt little by little, identities are complex and regions are made up of various strata of history and identity: how should this complexity be reconstructed? Where populations have different perceptions and different legitimacies in the same place, resilience strategies necessarily constitute choices that simplify and reorganize localities' identities. Resilience risks justifying a "return to a previous situation" which, in reality, is ideologically constructed, simplified and, indeed, instrumentalized.

It is evident here that the reconstruction process is a slow process with many players. Resilience involves taking account of multiple players not only in the risk prevention phase, but also during the reconstruction of places and their complex identity. A real resilience process takes place when it manages to reconstruct a shared area, a community, an urban link; the process becomes an ideological and potentially controversial construction when it favors a single stratum of the palimpsest, when it erases one part of a place's history to favor another. We can refer again to the example of Auschwitz being presented as a concentration camp for communists until the 1980s.

10.3.2. *The example of Dresden*

Chloe Voisin has studied Dresden at length, a historic city [CHL 09], 80% of which was destroyed at the end of World War II, reconstructed, and made part of German heritage. Dresden was recently excluded from the UNESCO World Heritage List; this exclusion, following construction on the banks of the Elbe and therefore indirectly linked to the historic reconstruction of the city center, is symptomatic of a specific attitude to heritage, a retreat from the criteria recognized by the international community. Voisin speaks of a "digestion" time for this city: the digestion of the different ideologies that have marked different phases of the city's reconstruction. She notes that the history of reconstruction is similar to a history of destruction (of material and identity), following several stages. First, following the bombardments, the clearing of the center was undoubtedly excessive and created a sort of "blank page", in 1946, a pretext devised by the authorities in order to reconstruct a Soviet city stripped of its past, and a future showcase for an ideological system. In the course of the following decades, phases of planning, destruction and reconstruction alternated, affecting most of all the "former" city center, and conferring on it successively a Soviet, then a middle-class, German, western, baroque, modern identity etc. German reunification then gave rise to a desire for the city's "normalization" and "Europeanization" [VOI 12]. At this stage, the reins of political development were handed over to private enterprises, and the legacies of socialism were destroyed, such as the *Centrum Warenhaus* or the view from the Pragerstrasse, redeveloped and reduced in width.

Finally, is Dresden an example of a resilient city? The city's identity and history has been constantly revisited and reshaped for essentially political reasons. Can it be said that the urban system has been perpetuated, and has remained qualitatively the same? What memory, or more precisely what memories, are transmitted by the inhabitants of a city that is not merely traumatized but whose mourning has been chaotic and which today still seems to seek a stable path of identity? Perhaps it can be hazarded, at least, that Dresden has still not left a period of disruption, understood not as the historic era of bombardment, but a long phase of urban "storm" that has, through successive waves, made it very difficult to read its center's urban palimpsest. If resilience policies rely on adaptability and great flexibility, the case of Dresden, among others, shows that the equilibrium between flexibility and stability is difficult to find, especially when the disruption has been violent and long-standing.

10.4. Conclusion

The resilience approach enables the subtlety and complexity of memorial messages to be underlined and suggests alternative routes to develop better practice appropriate to the variety of risks, places and individuals. Finally, it assumes that the process of memory permits the system to be continuous while integrating the variety of perceptions and reconstruction steps. This memory can be made spatial, indeed normalized, by political choices such as the conservation and social interpretation of traumatic ruins. However, this symbolic construction of disaster and risk entails significant cultural and political corollaries as reminders that resilience is never ideologically neutral.

Thinking in terms of resilience implies that we confront several dynamics of extreme complexity: the complexity of interactions between stakeholders and the articulations of scales, the multi-layered history and identity, the complexity of messages of memory and forgetting, and the complexity of political and ideological stakes. Memory is vital for resilience strategies; but memorial messages demand delicate handling, they require good coordination and in-depth attention to the complexity of individual situations.

10.5. Bibliography

[AND 86] ANDRÉ M., "Cathédrale de Reims", *Monuments Historiques*, vol. 145, pp. 113–114, 1986.

[BAI 96] BAILLY A., *Risques naturels, risques de société*, Économica, Paris, 1996.

[BEC 92] BECK U., *Risk Society: Towards a New Modernity*, Gage Publication, 1992.

[CHO 92] CHOAY F., *L'allégorie du patrimoine*, Seuil, Paris, 1992.

[COM 10] COMFORT L.K. *et al.*, "Resilience is the capacity of a social system [...] to adapt in a proactive manner and to recover from perceived disruptions [...] as out of the ordinary unexpected", in COMFORT L.K. *et al.* (eds), *Designing Resilience: Preparing for Extreme Events*, University of Pittsburgh Press, 2010.

[DOV 96] DOVERS S.R., HANDMER J.W., "A typology of resilience: rethinking institutions for sustainable development" *Industrial & Environmental Crisis Quarterly*, vol. 9, no. 4, pp. 482–511, 1996.

[FAB 86] FABBRO S. (ed.), *1976-1986. La ricostruzione del Friuli. Realizzazioni, trasformazioni, apprendimenti, prospettive. Un approccio multidisciplinare*, IRES Friuli-Venezia Giulia, Udine, 1986.

[HAL 50] HALBWACHS M., *La mémoire collective*, PUF, Paris, 1950.

[JAC 05] JACKSON J.B., "De la nécessité des ruines", *De la nécessité des ruines et autres sujets*, Éditions du Linteau, Paris, 2005.

[JEU 01] JEUDY H.-P., *La machinerie patrimoniale*, Sens & Tonka, Paris, 2001.

[LAC 07] LACROIX S., *Ce que nous disent les ruines. La fonction critique des ruines*, L'Harmattan, Paris, 2007.

[LEB 10] LE BLANC A., "La conservation des ruines traumatiques, un marqueur ambigu de l'histoire urbaine" *L'Espace Géographique*, pp. 253–266. 2010.

[PRE 10] PRÉVÉLAKIS G., "L'amnésie contre la résilience. Charte d'Athènes, nettoyages ethniques et marketing urbain", in REGHEZZA M., DJAMENT G. (eds), monthly seminar, "Résilience urbaine" at the École normale supérieure, available at http://www.geographie.ens.fr/-Resilience-urbaine-.html, 2010.

[RHI 10] RHINARD M., SUNDELIUS B., "The limits of self-reliance: international cooperation as a source of resilience", in COMFORT L., BOIN A., DEMCHAK C. (eds), *Designing Resilience: Preparing for Extreme Events*, University of Pittsburgh Press, Pittsburgh, pp. 196–219, 2010.

[SIN 10] SINOU A., "Résilience et patrimoine de l'inhumanité", in REGHEZZA M., DJAMENT G. (eds), monthly seminar "Résilience urbaine" at the École normale supérieure, available at http://www.geographie.ens.fr/-Resilience-urbaine-.html, 2010.

[TIS 07] TISSERON S., *La résilience*, PUF, Paris, 2007.

[VOI 09] VOISIN C., "Reconstruire Dresde après 1990? Les enjeux patrimoniaux des projets urbains au centre-ville de Dresde depuis la réunification", in DJAMENT G. (ed.), monthly seminar "Politique culturelle et enjeux urbains" at the École normale supérieure, available at http://www.geographie.ens.fr/Archives-2009-2010.html, 2009.

11

Critique of Pure Resilience

As we have seen throughout this volume, resilience is buzzing to the point of becoming a victim of its own success. By being brandished, bargained and brewed, it has been morphed into a portmanteau word, borrowed for very diverse ends. Moreover, it is often linked to other notions in vogue (sustainability, governance, etc.) that show a similar plasticity. Resilience has both a fairly intuitive content, an eye-catching hook and an undecided long semantic train where everyone can rediscover their own meaning. Its multiple transfers across disciplines seem to legitimize its semantic and theoretical vagueness, to the point where resilience too often boils down to the promise of a bright new horizon.

Resilience is not a hardened concept: it is an open notion that discourses use above all to make themselves appealing. This is why we should refrain from speaking of it as a concept, or a pure concept. It is more an intuition – both attractive and elastic – a notion that is increasingly used both as a promise and as a threat. But it is no longer really a notion that stimulates reflection: it is an injunction that puts the reason to sleep. This shift from intuition to injunction makes it imperative to question the theoretical and political foundations of resilience. Resilience may prove to be toxic because any glittering promise discourages inquiry and any injunction implies a threat. It is

---

Chapter written by Samuel RUFAT.

precisely because resilience is an idea that seems attractive and powerful that it is compulsory to make an effort to cast a critical eye.

> *I have come to understand that confronted with a powerful explanation my first step is to wonder what it obscures. In a way, the more powerful the explanation, the more difficult it is to see what it obscures in the penumbra of its own light* [SAS 12].

11.1. Resilience to the test of discourses

Although urban disasters are numerous, history records only a handful of definitive disappearances [VAL 05]; examples include Angkor, Babylon or Tikal that perished in the distant past and in exotic latitudes [DIA 11]. It is therefore tempting to turn resilience into a new label, but resilience is not always for the best. First, attributing everything positive to resilience turns it into a self-fulfilling prophecy; for example, a city is only being resilient if it manages itself to get rid of all that is undesirable and overcome crises. Second, resilience pushes toward a reinterpretation of crises, disasters and the status of such victims. Finally, it is used as a political discourse that seems to shut down rather than to invite to debate. It is therefore necessary to make a resolute choice, from a critical perspective, to stop considering resilience as a "pure" concept in order to tackle its political and practical consequences.

11.1.1. *Resilience as a new label*

After having to assert their "sustainable" character, cities should next become "resilient". The multiplication of urban disasters and climate warming are the apocalypses that enable international institutions to bid cities to prepare, adapt and save themselves. Resilience has thus become a label, a gage of a region's quality using the certification of international institutions, enabling significant sums to be spent and numerous stakeholders to be involved, notably big business consortia, across public–private partnerships.

From year to year, the discourse of international institutions became more urgent as the tally of urban disasters continued to rise. These institutions thus changed their discourse as well as their method, passing from eschatology to expectancy and trading the stick for the carrot. Rather than bidding cities and regions to reduce their vulnerability, they chose to market resilience and so integrate their recommendations into the global competition of regions. Since emulation, innovation, regional marketing and city branding are the keys to success in the competition to attract capital, competence and commercial orders, the reinforcement of metropolises' territorial clout, that is to say their spatial role as a center of gravity, happens through enhancing their image, promoting their quality of life, and transforming them into a desirable product on an international scale. Labeling a compelling idea with a strong environmental connotation, such as resilience, is therefore a way of involving these themes in promoting a quality of life that is decisive in attracting businesses and managers. It was a brilliant idea: transforming risk management into an argument in the competition between regions and cities could potentially be a powerful lever for change. But cities and regions seem to indulge above all in an image contest, focusing their emulation on branding strategies rather than on regional development.

The United Nations (UN), for example, launched the campaign *Disaster Resilient City: My City Is Getting Ready* in 2009, as well as the labels *Resilient City Champion* and *Resilient City Role Model*, promoting good practice and the most promising projects. For its part, the World Bank put forward in 2009 the program *Climate Resilient Cities: A Primer on Reducing Vulnerabilities to Disasters* and drew up lists of good practice under the label *Climate Resilient City*. Similarly, in 2011, the European Union implemented the program *Transitioning towards Urban Resilience and Sustainability*, giving the "TURAS" label to 16 European cities. In France, in 2011, the Ministry for Ecology, Sustainable Development, Transport and Housing created the label *Gestion des Risques Territoriaux – Pour un territoire résilient* (Territorial Risk Management – For Resilient Regions), adopting the international norm ISO 31000.

From this starting point, public–private partnerships and research projects have multiplied in developed countries in order to harness the international funds made available for resilience [NEW 08]. Resilient cities' rankings, role models or good practices often bring out the same Northern metropolises, in particular global cities: Barcelona, Copenhagen, Stockholm, Vancouver, and above all Tokyo, New York, London and Paris. Thus, the city of New York has just adopted *PlaNYC 2030* by addressing urban resilience. *PlaNYC 2030* aims to "increase community resilience to climate and disaster risks" [PLA 12], according to the UN's own terminology. Yet the plan actually provides for infrastructure improvements, the planting of trees, volunteers repainting roofs white and improved information on risk, as well as insurance coverage for the population, all measures that were in place well before the label.

11.1.2. *Resilience between risk production and risk construction*

This new label seems to take note of the multiplication of urban disasters and a degree of failure in vulnerability approaches, even if – paradoxically – it does not question earlier management policies. In New York, after hurricanes Irene in 2011 (40 deaths and $7 billion worth of damage) and Sandy in 2012 (130 deaths and $80 billion worth of damage), the projects actually consist of accelerating the construction of three giant embankments to close off the bay [GOU 13] and consequently increasing the sense of safety rather than inventing the city's adaptation to the effects of climate change. The label "resilient", therefore, seems to be more a change in discourse rather than a change in policy.

Cities and urban centers concentrate a triple ideal of protection, rationality and environment control. However, urban risks have an endogenous character that can be interpreted through two prisms. The first prism stresses the production of risk by urban processes: risks are presented as a "social product" [BLA 94] that exposes dysfunctions in urban systems [DER 96]. Megacities have thus been described as "crucibles of hazard" due to the interactions between urban growth, hazards and vulnerability [MIT 99]. The second prism presents risk as an artifact at the heart of modernity [BEC 86] – a "social construct" – that enables individuals to transform dangers and uncertainties into conjectures to guide their behavior.

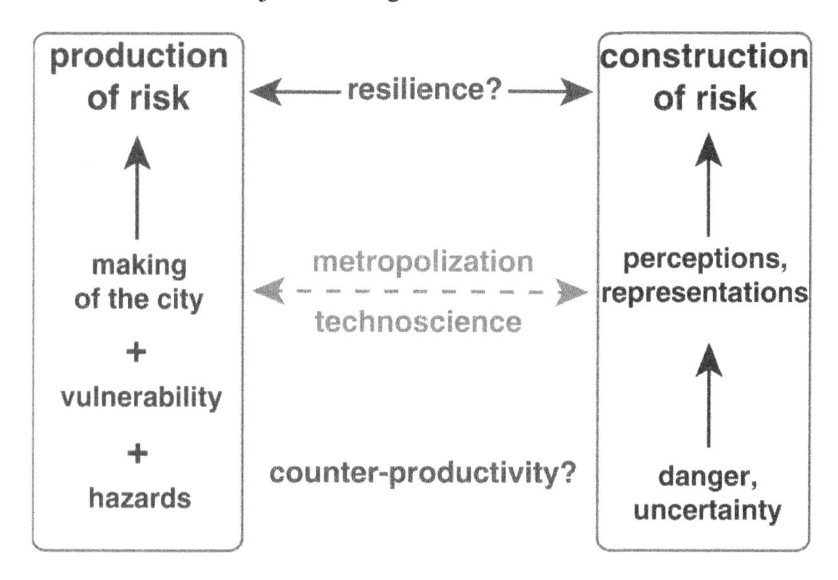

Figure 11.1. *Two approaches of risk in social sciences*

These analyses reveal a paradox (Figure 11.1): the need for security and forecasting, in order to guide decisions, ends by creating general uncertainty. The predictable retreats like a horizon as we try to approach it because the number of factors to be taken into account increases in line with our level of knowledge and even further as our capacity to store and handle data increases. Stakeholders are obliged to consider their actions on the basis of knowledge which they know to be incomplete, while the multiplication of databases on urban regions probably results in an overestimation of urban risks. Articulation between the production and construction of risk thus arises from urban "counter-productivity", as Illich [ILL 09] interprets it; for example, this renders a city's most impoverished inhabitants captive by reinforcing their vulnerability in place of offering them the protection they came searching for.

These apparent contradictions are adverse effects that remain masked because the means and ends collide. This "counter-productivity" in urban areas is in opposition to the current horizon of sustainability and urban resilience which has to remove adverse effects. The paradigm is still aiming to reduce risks and vulnerability.

11.1.3. *The waltz of notions*

These questions are inevitable because all-encompassing notions such as resilience can rapidly become counterproductive. It is always tempting to grasp an attractive idea and enlarge its field of application in order to take it from one discipline to another and test its heuristic potential until it is exhausted. We have seen in the first part of this book that the transfer of resilience from physics and ecology to the social sciences is not always accompanied by a solid theoretical base and that these issues are doubled by difficulties in establishing appropriate criteria for analyzing resilience and making it usable in risk management.

Discord even seems inevitable because ecological frameworks cannot be applied directly to societies: societies or cities are simply not ecosystems. Even economists have eventually recognized that cities do not always behave like ecosystems: a city's metabolism indeed follows the same power laws as all biological organisms, but this is absolutely not the case with social relationships and, in particular, not with innovation which is at the heart of urban development [GLA 11]. The urban economy is equipped with stakeholders and institutions that hold ethical values, unlike ecosystems. It is driven by individuals and groups that are capable of developing a reflexive approach to their situation and actions [FOL 09]. Because of this, the notions of "boucing back", "return to equilibrium" or "reorganization after disturbance" that resilience has inherited from ecology, depend on the social organization and above all on the viewpoints of the stakeholders [DUI 10].

Speaking of resilience in cities, it is therefore sensible to recall an obvious fact: societies are neither metals that undergo distortions without a word, nor ecosystems whose entropy and reaction to stimuli can be measured while ignoring political options and social consequences.

This rapid expansion of the field for applying resilience and the related disagreements have fueled polysemy, if not cacophony. Thus, when the most impoverished are the first to return to live in their devastated districts after a disaster, some will assess this as a sign of their resilience, others as a sign of their vulnerability, others of the metropolitan system's inertia or indeed its populations' captivity, without any of these categories necessarily corresponding to the population's experience. In the previous sections, we have seen the theoretical and practical implications of this melee of terms and abundance of interpretations. The situation is even more equivocal as some of the notions arise from observation, such as a crisis or vulnerability, and others from anticipated projects or horizons, such as resilience. The epistemological and political implications remain to be analyzed.

Now, it seems no longer possible to avoid disasters, and it is therefore necessary to adapt to them: this is what resilience appears to mean. But a look back at work on risks shows that notions succeed one another as they encounter obstacles to their use. Thus, the geography of risk has long given priority to hazards and dangers [MOR 06]. Then, faced with the inability of science and technology to eradicate threats, societies' ability to adapt has come to the fore [BUR 93]. Geographers then used vulnerability to "de-nature" disasters [WIS 76]. To avoid too passive a reading of societies' role, they then used resilience to mean the ability to overcome a crisis and adapt to it [PEL 03]. Just like reconstruction and adaptation, resilience enables a more positive approach to risks and disasters to be found [FOL 06]. Vulnerability appears to derive more from the point of view of risk production and a collective approach and develops before crises, whereas resilience seems to arise more from the point of view of risk construction, an individual approach and the aftermath of crises, over longer temporalities. But whereas vulnerability was linked to critical approaches, revealing links between disasters, poverty and development [HEW 83], resilience on the contrary offers a consensual and integrative project [LAL 08].

However, these notions are most effective for carrying out studies after the event. It is difficult to make them usable and they seem to condemn us to wait for disasters to occur to enable the science to progress.

This passage from one notion to another, resulting from difficulties in structuring them, thus seems to expose an avoidance of the problem. Vulnerability is a property independent of hazards and crises and can be studied with the aim of preventing disasters, whereas resilience requires reference to a disaster, to a shock. This is because a society or region that is vulnerable will undergo crises and will find itself in a position to adapt and derive lessons from the disaster. Faced with these difficulties, some researchers consider resilience too vague to be used in risk reduction strategies [MAN 06].

11.1.4. *Resilience as a political discourse*

In reality, resilience refers implicitly to normative questions, what a "good" city, a "good" society and "good" inhabitants should be. [RUF 10] Resilience is, therefore, primarily a discourse, and a political one.

The discourse of resilience constructs a "good" city that should be perpetuated, as opposed to a "bad" city that must be corrected by seizing the opportunity urban disasters present. This discourse was a powerful driving force for the reconstruction of Chicago after the 1871 fire, even if the term was not used at that time [HAR 04]. The city center had been razed entirely, but the city seemed to recover from the fire stronger: entrepreneurs transformed the disaster into an opportunity, resulting from the release of real estate and a discourse on renaissance that was willingly entered into and culminated in the organization of the Universal Exhibition in 1886. Chicago was able to attract capital from across the United States and to invest it in the first skyscrapers. The "fireproof party" overthrew the outgoing mayor by promising to rebuild an invulnerable city, to turn Chicago into a phoenix rising from its flames.

In Chicago, resilience has a dimension of political instrumentalization, city-branding and *a posteriori* reconstruction.

More recently, in New Orleans (see Chapter 6), there is no longer a question of urban renaissance, that is to say resilience as a property, but of taking advantage of resilience as a process of mutation, as an opportunity to impose sometimes radical transformation. After hurricane Katrina, politicians and business owners have once again mobilized the resilience discourse to present disaster as an opportunity:

> *I think we have a clean sheet to start again. And with that clean sheet we have some very big opportunities (...) Most New Orleans schools are in ruins. This is a tragedy. It is also an opportunity to radically reform the educational system and to convert public schools into charter schools. [...] We finally cleaned up public housing in New Orleans. We couldn't do it, but God did it* [KLE 08].

The reconstruction of New Orleans has enabled rapid returns on investment. This influx of capital has led to a radical transformation, with the dismantling of social housing and public services [DAV 10]. New Orleans has thus changed from a city of poor tenants to a city of well-off property owners [HER 10]. Business owners have rebuilt a "good" city, white, rich and deregulated on top of a "bad" city, black, poor and dependent of federal transfers. And the resilience discourse has also enabled politicians and entrepreneurs to be relieved of responsibility, as much for the disaster as for the choice of this radical transformation.

Beyond its immediate meaning, close to adaptation and reconstruction, each author, each stakeholder, understands resilience in their own sense, depending on their culture, sponsors and agenda. And whatever perspective and interpretation of resilience is accepted, it *is always said by a third party*. The precise moment someone tells the resilience, the status of the speaker and the criteria used to establish it have therefore to be carefully scrutinized. Staging as quickly as possible the reconstruction's success is a striking political act, aggrandizing the city and its leaders. On the contrary, not speaking of resilience can enable governors to maintain exemption situations, to designate guilty parties or to request aid.

In fact, whether resilience is named *a posteriori* as the administrative or scientific report of a process, or indeed shaped *a priori* as a project to improve some of a city's properties, it is first a political discourse. It is therefore crucial to know whether the discourse of resilience is formulated around a democratically chosen public project, or indeed if resilience is recommended by institutions that, more or less explicitly, coercively or using incentives, impose it on societies and individuals.

Even if the notion of resilience is distanced from any idea of turning back the clock or returning to an initial state, it remains a concept of ending crises through a return to normal. This is what gives the resilience discourse a strong normative and political charge: affirming that there has been a return to "normal" implicitly supposes that normal can be defined. Behind resilience, there are first political choices at play concerning the function of societies. Resilience's problem is not so much creating a new urban utopia, a vague promise of alluring tomorrows; it is creating a normative narrative, to impose a particular conception of societies and cities implicitly, using both incentives and coercion. This large gap between a semantic flow on the one side and normative discourses on the other makes us think that there is a serious risk in using this notion. Resilience then possesses a dark side largely masked by its recursive promise of bright horizons.

11.2. The dark side of resilience

At the Association of American Geographers' Annual Meeting in 2012, numerous sessions on resilience were organized in New York, in front of more than 8,000 people from across the world, and the conclusion was unanimous: it is necessary to speak of resilience "because it sells well".

Perhaps there is an opportunity to be seized here, but it is still necessary to know what is being sold when we speak of resilience. In the absence of any convergence on a stable definition and shared criteria, the notion of resilience is not neutral. Therefore, it seems imperative to take time to consider what this fashionable term conveys and the reasons behind the influx of capital that its popularity has generated. To attribute everything positive to resilience and everything negative to vulnerability is a quick way of promising better tomorrows. But from what do these discourses deflect attention?

11.2.1. *The fallacies of resilience*

Just like sustainable development, if resilience is necessarily "good", it is an injunction and not a notion that stimulates reflection. It often leads to a binary logic: attributing everything positive to resilience and everything negative to vulnerability, which is then reduced to the idea of fragility [CUT 06]. Resilience then seems to be a horizon of expectation and a post-disaster mantra narrative [LAL 08]. But it can only be validated after the event, as a reinterpretation of different stakeholders' experiences. It is, therefore, simultaneously a category for interpretation and a legitimization, through expertise, of decisions aiming to impose "good" choices on inhabitants by brandishing the threat of disaster. It is also an opportunity for engineers to dress up their often disparaged technical fixes with the now attractive image of resilience and sustainability.

Still, resilience has not resolved the problems on which the risk management concepts it supplanted also faltered. Adaptive capacity and vulnerability stumbled on methodological pitfalls when attempts were made to establish operational levers. These concepts have been pulled between an analytical approach that reduces their complexity without managing to include all the relevant factors and a synthetic approach that condemns them to retrospective analyses.

This is also the case for resilience. And it is surprising that it was derailed by the same obstacles, which indeed seems to confirm that the emphasis on resilience has foundations that are much more political than scientific. The phrase "there is no zero risk" can be applied to vulnerability, as well as to resilience. Indeed, there is no invulnerable society, city or region. Moreover, the focus on cycles of destruction/reconstruction makes all cities appear resilient. But resilience supposes in the first place that the urban fabric has been affected, some inhabitants have been killed and cities have lost something. But what is lost in resilience? And why is it never spoken of?

11.2.2. *The resilience imperative and social Darwinism*

The promotion of resilience does not necessarily reflect a transition from a negative to a positive approach. Vulnerability, which is always to be reduced, integrates resistance, the absorption of shock and adaptation. It does not assume that societies are passive; on the contrary, it is about analyzing the choices and dysfunctions that lead to crises. Moreover, by presupposing that the most vulnerable societies need help to be able to face crises, vulnerability fitted well with international institutions' ambitious policies.

For its part, resilience does not imply active resistance by societies: on the contrary, it supposes they are fragile. Speaking of resilience infers that, in the first instance, the urban fabric and the functioning of cities and their inhabitants are affected, altered and crippled. Despite resilience's promises, cities do indeed lose something during the course of a disaster, and what they lose are primarily the most fragile buildings, the least adapted networks and their poorest inhabitants. Briefly, resilience conveys the idea that the most vulnerable can be swallowed up by a crisis, provided that the rest draw lessons from it and so seek to adapt. So making social Darwinism appealing is the dark side of resilience.

It is not merely a suppressed subtext, and Japan is a sad example of it. After the disaster at the nuclear plant at Fukushima in March 2011, the operator Tepco approached subcontractors, requesting them to "liquidize" the plant's site. These subcontractors then turned to the Yakuzas, the Japanese mafia, who recruited the most vulnerable and least aware of radioactivity, the unemployed, homeless, those in debt, etc. The most vulnerable were, therefore, those sacrificed and sent to the center of the radioactive exclusion zone. To stabilize the situation, by going to the heart of the reactors that had melted, looking for bodies, clearing rubble and starting a concrete sarcophagus, they were forced to receive – in 1 h – the internationally established annual maximal dose of radioactivity [SUZ 11]. Everything happened as if the sacrifice of the most vulnerable was a way of finding a social use for them (finally! some might perhaps say) by ensuring the rest of society's survival and eventually their adaptation. Resilience carries with it the seeds of social Darwinism and its use tends to normalize it, to impose it as obvious, indeed to make it appealing. If resilience is so toxic, it is because it has in its turn been thoughtlessly transposed from ecosystems to societies.

11.2.3. *Adapt or perish!*

The recent shift in discourses from vulnerability to resilience thus enables a radical change of approach in risk management to be glimpsed: societies' vulnerability, which is largely experienced by the poorest but which can be anticipated and reduced by aid systems dependent on collective solidarity and state involvement, is opposed to resilience, which is sought after, but can only be validated retrospectively and which leaves adaptation to the individual. Resilience thus becomes an injunction: adapt or perish!

In fact, since the beginning of the 1990s, the UN, the World Bank, the European Union and various regional organizations have developed a responsibility model based on individual failings [REV 09a]. During this period, these stakeholders developed an arsenal of programs, norms or guides that indicate how to overcome a disaster "well" and above all how to prepare to lessen the consequences. At the beginning of the 1990s, international organizations relied on vulnerability to target the physical and social characteristics that called for a responsibility to take charge. The program shifted rapidly from causes to individuals with the figure of the pure victim emerging, incapable of recovering independently from the crisis that they have undergone [REV 09b].

Then, faced with budgetary restrictions and the increasing cost of disasters, these programs were accused of inducing passivity on an individual level. Agencies, therefore, introduced the notions of participation and individuals' coping ability, and they highlighted a model of responsibility that emphasized individual involvement [AMB 09]. The imperative to adapt to face climate change is clearly formulated in the UN's 2004 report *Living with Risk*.

The notion of resilience was then highlighted as a more supple strategy with the adoption of the Hyogo Framework for Action in 2007. Finally, the program *Making Cities Resilient*, launched in 2009 by the UN, relies on infrastructure and public–private partnerships to develop technical solutions, educate populations and enable the individual to take responsibility. Local communities are enjoined to take responsibility for themselves on pain of disappearance: making stakeholders responsible results in those who do not meet "good" practice being held culpable. It constitutes a new reading of disasters that enables the most impoverished to be stigmatized; they are no longer considered to be victims but actors who should educate, motivate and organize themselves to act. These discourses of injunction to resilience could lead to individual responsibility being imposed on individuals affected by crises, regardless of negative interactions in their social milieu or environment, or the possible benefits of the crisis for the rest of the society.

11.2.4. *The return of a moral interpretation of disasters*

The promotion of resilience also marks the resurgence of moralizing discourses, with very acerbic criticisms of actors who do not conform to the dominant model. Resilience understood as a property enables a sort of teleology to be postulated and justifies a form of disengagement by traditional players in disaster management.

Understanding resilience as a process leads to adaptation being imperative. On an individual level, this injunction translates into social Darwinism. The change from vulnerability to resilience leads to a change in political strategy: we pass from planned prevention to an emphasis on the local and the role of communities and to individuals being made responsible. At a collective level, it smoothes a renewal of the moral interpretation of crises and disasters, passing from the heroification of some stakeholders, some cities and some regions to holding those who do not meet "good" practices culpable.

This reading of disasters, in fact, permits those who do not take part in the adaptation process to be stigmatized, without, however, examining the causes behind their passivity or their resistance. Disasters can then be presented teleologically as an opportunity for a "necessary" purification of "bad" (vulnerable) cities so that "good" (resilient) cities can emerge after the disaster. These mechanisms were observed well before resilience was introduced in discourses, but resilience made the idea of a "normal" functioning of cities and societies fashionable again.

In these conditions, the vogue for resilience quite simply suggests a return to the 18th Century and a conception of disasters as "divine punishment". By using a horizon of expectation desirable to everyone, resilience enables choices to be imposed, a "good" city, "good" citizens, the "deserving (good) poor", return to a "good" previous state, etc. Resilience thus belongs to an understanding of disasters that is both moral and teleological, with a linear approach to time, tending toward societies' progress or adaptation, as opposed to a cyclical timescale where crises and disasters return. This is what prevents it from being seen that resilience processes can also lead to damaging situations being maintained or counterproductive processes being used again. This linear approach explains international programs' insistence on an injunction to adaptation. And this moral dimension is accompanied by a change in the individuals they address, who are no longer victims but actors.

The UN's campaigns reveal this change of discourse well: the iconography has changed: up until 2005 the archetype of vulnerability in the International Strategy for Disaster Reduction (ISDR) brochures was a black woman in distress, alone or with a child in her arms. Since *Towards resilient cities*, it has been replaced by an archetype of resilience: muscular, smiling men in the middle of rebuilding a dispensary or dam. These discourses are much removed from the radical reading of the 1970s that had introduced the notion of vulnerability into international programs to denounce the economic and structural causes of disasters. Since then, and equally far from the

Millennium Development Goals, it has been a question of glorifying heroes who emerge (by themselves, if possible) after disasters. At the level of the city, the same heroification can be observed: model cities, good practices and local leaders are showcased as champions of resilience, etc.

Management policies consider victims as citizens, struck individually by disasters and misfortunes, that public authorities take charge of for humanitarian reasons. They are ordered to take part in their moral and social rehabilitation and to invest in action taken to help them by using their own resources. This method of governing the most vulnerable relies on the ancient law that compensation is expected from a citizen to cover the assistance he receives. He has the moral or civic duty to return what the state gives him in the form demanded of him, thus proving his desire to *pull through* [THO 10]. The injunction to resilience could therefore be analyzed as a new means of governing the vulnerable.

The consequences of this injunction can also be found at international level, as the example of Haiti after the January 2010 earthquake shows [COM 10]. The media was saturated by images such as that of the inhabitants of Port-au-Prince returning to pray in their ruined churches surrounded by bodies or the strength of character of the little girl who had lost all her family and her legs, whereas others emphasized the fatalism that condemned the Haitians to resign themselves to chaos [HUT 11]. Thus, it was the discourse of resilience that was mobilized, and not Jared Diamond's analyses, which had shown that the colonial powers have perpetuated the isolation and vulnerability of Haitian society for nearly two centuries [DIA 11]. For the disaster at Port-au-Prince is not to be sought in the "qualities" of the Haitians but in the island's underdevelopment, something the discourse of resilience tends to make one forget.

11.2.5. *"There is no alternative"*

We have shown that resilience has a toxic power because, if the notion is pushed to its limits, it renders social Darwinism desirable, imposes a moral reading of disasters and essentializes vulnerability, contrary to critical approaches that led to international solidarity and aid. These analytical implications of resilience open up terrifying possibilities. This does not mean, of course, that all stakeholders use it *voluntarily* to excuse, justify or legitimize the most hypocritical and egoistical discourses, the most barbarous conceptions and the most ignoble choices. However, resilience also presents another phantom menace because its elasticity favors its instrumentalization.

Discourses of resilience permit some stakeholders to dismiss all alternatives by alternating between the threat of cataclysms and the promise of a glorious future. "*There is no alternative*" – just as in the era of Margaret Thatcher and Ronald Reagan, some political, economic and financial players seemed to be seeking to impose a program by refusing to debate choices that are critical for society. The discourses of crisis now spill over into the economy; they instrumentalize resilience to make choices seem natural and to make neoliberal paradigms and experts, accountants and engineers fixes.

Programs devised by international institutions and the branding of resilience lead unerringly to technical and technological fixes, to investment, to the market and to public–private partnerships. These solutions make one think, once again, of the scientist and technicist myths, after the fusion of the reactors at the Fukushima nuclear plant in 2011, after the sinking of the platform Deepwater Horizon in 2010, after the disastrous technical and organizational failures brought to light by Xynthia in 2010, the Sichuan earthquake in 2008, hurricane Katrina in 2005, etc. This time, international campaigns argue that networks will become intelligent, energy neutral, cities green, external factors positive, etc., forgetting that the 20th Century showed us that if technical progress enabled some risks to be lessened, it symmetrically created others, multiplying our destructive power [BEC 86]. It seems fairly naive to think that, overnight, for the first time in human history, technology will henceforth cease to produce undesired side effects, and moreover will remedy all the problems it previously caused [DIA 11].

Moreover, discourses of resilience are also useful for promoting deregulation, responsibility and competition in individuals, according to the well-established recipes of free market supporters:

> *The path back is long and hard. Cities must return to their roots as place of small-scale entrepreneurship and commerce. Apart from investing in education and maintaining core public services with moderate taxes and regulations, governments can do little to speed this process. Some places will, however, be left behind. Not every city will come back, but human creativity and entrepreneurial innovation are very strong (...) Cities like Rio have plenty of poor people, because they're relatively good places to be poor. After all, even without cash, you can still enjoy Ipanema Beach* [GLA 11].

It was Milton Friedman himself who, in a column in the *Wall Street Journal* on 13 September 2005, two weeks after Katrina, championed his recommendations for facilitating the reconstruction of New Orleans: to make it a free zone, to temporarily abolish labour laws, to relax environmental regulations, etc., in order to promote the resumption of economic activity. For neoliberals, budgetary restrictions and the increasing cost of disasters justify an increasing recourse to the world of business. Indeed, in the months that followed Katrina, the federal government distributed more than $3 billion to multinational companies, without competition procedure, in the form of a contract to distribute humanitarian aid, clearance of rubble and reconstruction [KLE 08].

In reality, Katrina served to reveal a privatization of risk management, the substitution of businesses for public authorities and states in disaster management. The Red Cross, for example, signed a contract with Walmart, declaring that the business had more resources, better logistics and critical expertise [ROS 11]. In New Orleans, the discourse of resilience permitted multinational businesses close to the ruling powers to install an economy of plunder that occurred through the exploitation of the most vulnerable. The poorest of those affected were abandoned to their fate, while reconstruction was directed first toward property owners and solvent inhabitants: businesses responded to the destruction of social housing by constructing a city of private landlords, private schools and private hospitals [HER 10]. Companies took advantage of the resources and the deregulation made available by the federal authorities to charge aid and reconstruction missions a high price. The federal authorities used the discourse of resilience to promote their steps to stimulate the economy and revitalize regions. But behind the discourses were the multinational businesses, often those who had earlier participated in the reconstruction in Iraq, who crushed local businesses. Moreover, they called on immigrant labor, often unregistered, which enabled them not to pay their employees, rather than employing the local population. After Katrina, the local population was thus condemned to idleness, and those affected to watch impotently as businesses and workers from outside destroyed social housing and public services, exporting the profits elsewhere [DAV 06].

This privatization of risk management, this substituting of businesses for public authorities and states in managing disasters has reached such a point that world stakeholders in insurance have come to consider crises, reconstruction and resilience as new, trivial opportunities. They have even come to return to the notion of a "moral hazard", which is usually used to denounce the fact that insurance stakeholders and finance take unnecessary risks because they know that in case of crisis, governments will oblige taxpayers to rescue them. Indeed, after Katrina, insurance stakeholders went so far as to declare that governments were entering into unfair competition with them, because if people are confident that they will be helped "for free" and aided in case of disaster, they have no incentive to take out private insurance:

> *The compassionate federal impulse to provide emergency assistance to victims of disasters affects the market's approach to managing its exposure to risk* [KLE 08].

A division was created between those who could pay for their own insurance and resilience and those who were victims of both the disasters and the budgetary restrictions on missions governing aid and crisis management. The stakeholders in this new market left no doubt. Behind their discourses of resilience, disasters only affect insolvent individuals; for the others, it is always possible to turn them into an opportunity for a luxury holiday:

> *The first hurricane-escape plan that turns a hurricane evacuation into a jet-setter holiday at a five-star golf resort. No standing lines, no hassle with crowds, just a first class experience that turns a problem into a vacation ... Enjoy the feeling of avoiding the usual hurricane evacuation nightmare* [HEL 12].

This was already perceptible after the December 2004 tsunami in Thailand or Sri Lanka, after the destruction, where the programs aiming to enable traditional fishing communities to recover had served as a pretext to displace these populations and distribute their land to property consortiums who took advantage of the situation to develop new luxury tourist complexes [KLE 08].

Consequently, resilience conveys ideas of individual responsibility, competition and opportunity. This is what facilitates its use for deregulation, privatization and, finally, the transformation of disasters into a new markets. Discourses of resilience are an opportunity to designate the victims as primarily responsible for crises and disasters and then to exalt political voluntarism and market competition as the solutions. This use enables a modification in urban functioning to be imposed through deregulation and opening new markets. Discourses of resilience are thus used by some stakeholders as rhetorical tools to justify choices that open up an urban *apartheid* in conjunction with environmental injustices and the fragmentation of the city. It is therefore essential to know who tells resilience, what kind of resilience, and above all, why.

11.3. "Good" or "bad", who is declaring resilience?

It is not enough to decree resilience; like adaptation, it requires a profound change in the way cities function. However, we have seen that it is first a discursive and normative instrument, a political discourse in the form of an injunction. Therefore, the only way of consolidating resilience and making it operational is to try to extricate it from its implicit nature, from preconceptions about the "good" city and the catalog of "good" practices. Briefly, to save resilience, it is paradoxically mandatory to wonder what "bad" resilience would be. Because those directly involved are never those declaring the resilience or spontaneously appropriating the notion of resilience; it is a discourse that is always brought from outside and which moreover translates (often implicitly) into moral judgments and normative injunctions.

11.3.1. *The resilience of slums*

Let us start from an observation – shantytowns are the default of urban disasters:

> *Slums begin with disasters. Precisely because the site is so hazardous and unattractive, it offers a protection from rising land values in the city. They are poverty's niche in the ecology of the city, and very poor people have little choice but to live with disaster* [DAV 06].

Slums are also the ordinary aftermath of a disaster. The face that cities present after major disasters shows that the most resilient form is precisely the slum, the most resilient process is the informal economy and the most resilient form of organization is the seizure of power by gangs. This was the case in Port-au-Prince after the 2010 earthquake, in Beirut after the war in 2008, and also in New Orleans after Katrina, and to some extent in the camps for Fukushima refugees, etc. We do not want to see it, revealing that resilience is, in fact, at least as much (if not more) a political instrument as a scientific notion: smoke and mirrors blinding us rather than enlightening us.

According to the UN Habitat program's definition, a population inhabits a slum or non-durable housing if it does not have access to drinking water or sanitation, if it does not have sufficient space per person, if it lives in a temporary structure or does not enjoy the security of permanent residence. Unofficial or self-constructed neighborhoods departed from utility networks are intrinsically the structure that can function most easily in a degraded state, even if they are no more than a pile of rubble or detritus, and which regrow most rapidly after a disaster, using light materials, without a need for infrastructure and without having to rule on the thorny question of property. Whichever the resilient city definition retained, one that is resistant, one that supports functioning in a degraded state or facilitates rapid reconstruction, it leads first and foremost to slums.

Slums are thus the dark side of resilience. It is a very different image from the enthusiastic discourses and the regional marketing and city-branding promoting resilience as the very figure of the future for cities finally reconciled to their environment and their inhabitants, a horizon of expectation for all to seek. But then a worrying discrepancy exists between the "good" resilience of the discourses, the resilience that governments and international bodies turn into an injunction and entrepreneurs turn into opportunity, and the "bad" resilience, which is overlooked even though it is the result of urban disaster.

The shift of management policies to resilience can therefore be interpreted otherwise. Vulnerability was consistent with the policies of international organizations by emphasizing the necessary assistance to the most vulnerable societies in disaster reduction. Whereas resilience, on the other hand, seems to be more in line with the current framework of deregulation, placing individuals and regions in competition with one another, of budgetary restrictions and of withdrawal of the states and international bodies.

We are not prepared to look resilience in the face when there is no one stating, declaring or admitting "bad" resilience. This "bad" resilience has no spokesperson, no opinion multipliers; it is never making the headlines, and therefore seems quite simply to have never existed. It is almost invisible because this is a matter of the daily, the ordinary city and the handiwork in the face of adversity more than the spectacular side of crises and disasters. Once value judgments, injunctions and bias toward the "good" city are removed, there is no choice but to acknowledge that the burgeoning of the informal sector is the most resilient urban process, and that slums are the most resilient urban form.

11.3.2. *Resilience and governmentality*

This reluctance to recognize slums' resilience explains the difficulties experienced in formalizing this notion, despite its promises. The paradox of resilience has to be torn between its bright horizon catchline and its basis as Darwinist injunction and slums endorsement. This paradox is the symptom that its use arises from city and regional branding, and is consistent with the Foucauldian concept of "governmentality" [FOU 04]. Governmentality is the art of using expertise and knowledge to govern minds and shape behavior both using incentives and coercion: it is the whole set of procedures, analyses and reflection, calculations and tactics allowing the rational rule of entire populations, of their actions and behaviors by making them internalize norms through knowledge and safety features.

Resilience can thus be read as a political discourse that fits a vertical mode of governing minds and behavior. This surely explains, behind the promises of security and triumphant horizons, the rapid shift from intuition to injunction. We must not let ourselves be blinded by the promises of resilience.

For more than 20 years, social sciences have criticized risk as a power strategy and the use of disasters to impose individuals to become "good citizens". This is one of the main reasons for the success of the concept of vulnerability. Radical approaches of risk, its representations and its management have used vulnerability to consider risk as a government tool, an injunction to internalize norms. The current shift from vulnerability to resilience seems to be bouncing back. It seems that we have let ourselves be imposed unreservedly a concept arising from the material and ecological sciences, then reinterpreted by self-development experts. Are we going to desist from any criticism of the political dimension of risk?

Watching resilience percolate into public discourse reveals a strange paradox in the current era dominated by a series of environmental, geopolitical, financial, economic and social crises that discredit decision-making bodies. States are participating in the move toward resilience – at the very moment they are divesting themselves of their basic infrastructures and reducing their spending, or even undermining the very notion of public service. However, two decades of research and lessons learned have established that they are instrumental investments for vulnerability reduction and disaster mitigation [MIT 99]. In fact, a way to explain this apparent paradox has been provided by the UN, itself putting forward the idea of "community resilience". This is easy, cheap and open to all: it means you only have to let each "community" manage just as it can afford or as it sees fit to become resilient, while still finding a justification for budgetary cuts and the current privatization of disaster management.

The borrowing of resilience from materials science should have alerted us to the risk for the social sciences of seizing on materialist, scientist and positivist conceptions without critical distance. The Haitian earthquake is a bitter reminder that disasters do not have a purifying role and that societies do not necessarily progress after a severe crisis. Political dysfunction, social inequality and hunger have a part in resilience processes. Therefore, resilience should not be an absolute horizon of expectation and it should not be sought only in the fabric, landscape and material structures of the city. Resilience involves an element of forgetting, which enables reconstruction, even identical reconstruction, and an element of adaptation, which imposes a change in urban structures but above all in urban functioning.

11.4. Conclusion

The key issue of resilience is consequently knowing who is declaring the resilience, what kind of resilience and why. It primarily appears as a post-disaster narrative mantra aiming to impose a "good" city on its inhabitants by raising the spectre of disasters. But it can only be validated *a posteriori*, as a reinterpretation of the different actors' experiences. It is therefore both an interpretation and a legitimating strategy. It is also an opportunity for stakeholders knowing how to dress up their technical fixes in the currently attractive image of a city that is both sustainable and resilient. In fact, the main impact is to strip regions and metropolises of any political dimension and to let citizens on the back burner.

Resilience is primarily a narrative, a discursive setting that appears to be a power strategy. This discourse is consistent with governmentality by using both a desirable horizon of expectation for all and the threat of disasters to come, resilience requires choices (the "good" city) and behaviors ("good" citizens and "good" practices) to be imposed. Moreover, by making the "bouncing back" desirable, the elite have an opportunity to maintain the social and political *status quo*. By focusing on the conception of adaptation resulting from ecology and making it compulsory, they also have a chance to legitimize changes in urban functioning through competition and individual responsibility, which *in fine* justifies disaster management by urban apartheid.

Whether the different stakeholders are aware of this or not, and whether or not they are seeking to use it, the current use of resilience is both toxic and perilous. It is all the more dangerous as it enables political debate to be shut down relying on scientific legitimacy. It is therefore compulsory to become aware of this phantom menace and to adopt a critical perspective, to resolutely and firmly deconstruct the discourses and mechanisms allowing a switch from the uncritical use of a notion to an instrument of power, the legitimization of anything and everything, in particular the worst that societies can do.

While vulnerability is consistent with the field of spatial justice and the imperatives of cohesion and regulation, current discourses on resilience are closer to the neoliberal sphere with an injunction to competition and individual responsibility, alleging market rules, deregulation and technical fixes as the inevitable solutions. Therefore, resilience should be primarily analyzed as a political discourse that is easily used to impose choices that should at least be publicly discussed, whereas its use tends to detract attention from political and social processes onto econometrics and technical fixes. Rather than a promising horizon, current use of resilience thus presents a major risk.

11.5. Bibliography

[AMB 09] AMBROSETTI D., BUCHET DE NEUILLY Y., "Les organisations internationales au cœur des crises", *Culture et conflits*, vol. 75, no. 3, pp. 7–14, 2009.

[BEC 86] BECK U., *Risikogesellschaft: Auf dem Weg in eine andere Moderne*, Suhrkamp Verlag, 1986.

[BLA 94] BLAIKIE P. *et al.*, *At Risk: Natural Hazards, People's Vulnerability and Disasters*, Routledge, 1994.

[BUR 93] BURTON I., KATES R., WHITE G., *The Environment as Hazard*, 2nd ed., The Guilford Press, 1993.

[COM 10] COMFORT L., SICILIANO M., OKADA A., "Risque, résilience et reconstruction: le tremblement de terre haïtien", *Télescope*, vol. 16, no. 2, pp. 37–58, 2010.

[CUT 06] CUTTER S. (ed.), *Hazards, Vulnerability and Environmental Justice*, Routledge, 2006.

[DAV 10] DAVIS M., "Who is killing New Orleans?", *The Nation*, 10 April 2006.

[DAV 06] DAVIS M., *Planet of Slums*, Verso, New York, 2006.

[DER 96] D'ERCOLE R., THOURET J.-C., DOLLFUS O., "La vulnérabilité des sociétés et des espaces urbanisés", *Revue de géographie alpine*, vol. 82, no. 4, pp. 87–96, 1994.

[DIA 11] DIAMOND J., *Collapse: How Societies Choose to Fail or Survive*, Penguin Books, 2011.

[DUI 10] DUITA A., GALAZA V., "Governance, complexity, and resilience", *Global Environmental Change*, vol. 20, no. 3, pp. 363–546, 2010.

[FOL 06] FOLKE C., "Resilience. The emergence of a perspective for social-ecological systems analyses", *Global Environmental Change*, vol. 16, no. 3, pp. 253–267, 2006.

[FOL 09] FOLKE C., ROCKSTRÖM J., "Turbulent times", *Global Environmental Change*, vol. 19, pp. 1–3, 2009.

[FOU 04] FOUCAULT M., *Sécurité, Territoire, Population. Cours au Collège de France, 1977–1978*, Seuil, Paris, 2004.

[GLA 11] GLAESER E., *Triumph of the City*, Penguin Press, New York, 2011.

[GOU 13] GOULDEN M., ADGER N., ALLISON E. *et al.*, "Limits to resilience from livelihood diversification and social capital", *Annals of the Association of American Geographers*, vol. 103, no. 4, pp. 906–924, 2013.

[HAR 04] HARTER H., "Chicago et l'incendie de 1871: Entre mythes et réalité", in CABANTOUS A. (ed.), *Mythologies urbaines. Les villes entre histoire et imaginaire*, Presses Universitaires de Rennes, Rennes, pp. 219–236, 2004.

[HEL 12] HELP JET, available at www.helpjet.us, accessed in 2012.

[HER 10] HERNANDEZ J., ReNew Orleans? Résilience urbaine à la Nouvelle-Orléans après Katrina, PhD Thesis, University of Nanterre, 2010.

[HEW 83] HEWITT K., *Interpretation of the Calamity from the Viewpoint of Human Ecology*, Allen and Unwin, London, 1983.

[HUT 11] HUTTES C., "Hope for Haiti", *Illinois Wesleyan University Magazine*, vol. 20, no. 1, p. 2, 2011.

[ILL 09] ILLICH I., "Énergie et équité (1973)", *Œuvres complètes*, vol. 1, Fayard, Paris, pp. 399–402, 2009.

[KLE 08] KLEIN N., *The Shock Doctrine*, Penguin Books, London, 2008.

[LAL 08] LALLAU B., "La résilience, moyen et fin d'un développement durable?", *Ethics and Economics*, vol. 8, 2008.

[MAN 06] MANYENA S.B., "The concept of resilience revisited", *Disasters*, vol. 30, 2006.

[MIT 99] MITCHELL J.K., *Crucibles of Hazard: Mega Cities and Disasters in Transition*, Brooking Institution, 1999.

[MOR 06] MOREL V., DEBOUT P., "Regard rétrospectif sur l'étude des risques en géographie", *L'information géographique*, vol. 70, pp. 6–24, 2006.

[NEW 08] NEWMAN P., BEATLEY T., BOYER H., *Resilient Cities: Responding to Peak Oil and Climate Change*, Island Press, 2008.

[PLA 12] PlaNYC 2030, A Greener, Greater New York, available at http://www.nyc.gov/planyc, accessed in 2012.

[PEL 03] PELLING M., *The Vulnerability of Cities: Natural Disasters and Social Resilience*, Routledge, 2003.

[REV 09a] REVET S., “Les organisations internationales et la gestion des risques et des catastrophes ‘naturels’”, *Les Études du Ceri*, vol. 157, available at http://www.sciencespo.fr/ceri/sites/sciencespo.fr.ceri/files/etude157.pdf, September 2009.

[REV 09b] REVET S., “Vivre dans un monde plus sûr crises”, *Culture et conflits*, vol. 75, no. 3, pp. 33–51, 2009.

[ROS 11] ROSENMAN E., The road away from home: policy and power in post-Katrina New Orleans, PhD Thesis, University of British Columbia, available at: http://hdl.handle.net/2429/38039, 2011.

[RUF 10] RUFAT S., “Existe-t-il une ‘mauvaise’ résilience?”, *Séminaire Résilience Urbaine*, ENS Ulm, available at http://www.geographie.ens.fr/Compte-rendus-de-seances-2010-2011.html, accessed in 2012.

[SAS 12] SASSEN S., “Analytic tactics: geography as obstacle”, *AAG Honorary Geographer’s Lecture*, New York, available at http://www.aag.org/cs/annualmeeting/videos/2012, accessed in 2012.

[SUZ 11] SUZUKI T., Yakuza to genpatsu Fukushima Daiichi sennyuki, (The Yakuza and Nuclear Power: An Undercover Report from Fukushima) Tokyo, Bungei Shunju, 2011.

[THO 10] THOMAS H., *Les vulnerables. La démocratie contre les pauvres*, Éditions du Croquant, Paris, 2010.

[VAL 05] VALE L.J., CAMPANELLA J.T. (eds), *The Resilient City: How Modern Cities Recover from Disaster*, Oxford University Press, 2005.

[WIS 76] WISNER B., O’KEEFE P., WESTGATE K., “Taking the naturalness out of natural disaster”, *Nature*, vol. 260, no. 5552, pp. 566–567, 1976.

Conclusion

"In a world of complexity and contingency, of risk, rationality, flows and mutability, theoretical frameworks that promise a means of capturing that complexity are seductive" [WEL 13].

Resilience seems to be the ideal response to demands for reassuring futures, for certainties within contingencies, for promises in the face of dangers and for persistence in change. Yet this idealized recipe still fails the test when put into use, which makes its transition from a visionary intuition to a practical solution quite challenging. Resilience cannot be used simply as a toolbox: it involves taking account of the increasing complexity of an ever more interconnected world, where each action causes retroactions on different levels, in distant places and on uncertain timescales. This is what explains the difficulty in defining and formalizing it: faced with complexity, resilience seems condemned to elasticity.

The different contributions suggested shed light on the richness of resilience approaches and the challenges that remain to be overcome. This first overview enables many observations to be made and a number of directions for further thought to be put forward.

Resilience provides an undeniable heuristic interest for considering, on the one hand, the dialectic between stability and change and, on the other hand, the "complexity turn [URR 13]" of the social sciences in the context of uncertainty.

Resilience can, in fact, provide food for thought on the future of ecological, social or regional systems and enables a departure from a purely linear vision of their trajectory. By questioning the notions of equilibrium, norms and adaptation, it challenges change and instability by allowing differentiated rhythms, ruptures and transitions, as well as discontinuities, etc., to be considered in the long run.

The different contributions in this book emphasize that there is not one but several resiliences, as much within the social sciences as in cindynics or ecology. It also responds to the growing complexity of societal goals, which calls not only for cross-disciplinary insights but increasingly defies academics' efforts, as much in terms of definition as comprehension. In this context, the metaphorical use of the term "system" appears to be nothing more than a stop-gap measure for defining these goals, at least in discourse. More broadly, resilience illustrates the inherent challenge in an cross-disciplinary approach and in transferring ideas from one discipline to another. Mobilizing resilience, in fact, requires a prior effort at rigor and definition, not to fix a term whose polysemy creates richness but to make good use of this diversity, to emphasize the nuances and avoid misunderstandings. In this sense, studying resilience calls for reflection on how interdisciplinary research is practiced.

It is convenient here to return to the contributions from ecology. Even though scientific ecology has been home to numerous debates on the use and meaning of resilience, today there is a paradoxical reversal. Since its beginning, scientific ecology has acted as a reservoir of notions, methodological tools or theoretical frameworks for the social sciences. Borrowings and transfers have always occurred using an analogical method and with multiple precautions being taken. Ecology has notably enabled scholars to go beyond linear causality and mesological determinism to consider interactions systemically or to question the processes' multiple temporalities. In the domain of hazard and risk in particular, it has enabled scholars to de-naturalize disasters and to socialize and politicize them. Ecology has consequently fed into the social sciences. However, with resilience, the ecological reference has been twisted: the "great leap backward" at issue here indeed causes scholars to mobilize notions taken from ecology indiscriminately or without intellectual rigor, obscuring the methodological and theoretical debates in human and social ecology, as well as botanical ecology, such that questions that are primarily precisely social and political are re-biologized and de-politicized. Although the practical consequences have been clearly established, the methodological and scientific consequences are just as pernicious: although scholars in the social sciences will not want to abandon contributions from scientific ecology, it is necessary to reassert the conditions for borrowings and transfers between disciplines.

Resilience finally sheds light on the hybridizations between nature and culture, the interconnectedness of spatial and temporal scales, the interactions between social and spatial scalar levels, the retroactions between systems' and subsystems' different components, etc. Indeed, this complexity echoes a necessarily anxiety-provoking uncertainty. In this sense, "*Resilience discourses mark a break with the modernism of the 'risk society' by introducing novelty, adaptation, unpredictability, transformation, vulnerability* [WEL 13]". Undoubtedly, this is what explains the concept's success in the field of climate change, since this filed precisely defies social expectations on prediction abilities and environment control.

In these conditions, although resilience does not necessarily create new scientific or management paradigms, it is integral to a precise moment in the history of Western society, which some have described as the crisis of modernity, for which it is both a reflection and a response. In any case, it opens the way for fresh thought and action frameworks to be taken into account. For example, it can lead to "new" risk and disaster cultures being defined – the novelty being entirely relative here. Thus, in her work on Japan, M. Augendre [AUG 11] recalls that for the Japanese, "*catastrophe is not a fall or an outcome (cata-), but an anastrophe, a new direction, a new development (ana-); it is the point of departure for a necessary adaptation*". N. de Richemond has similarly shown how rural societies before the French Revolution, in a premodern context, adapted to disasters [MES 10] using processes that derive from the resilience processes much lauded today.

However, the promotion of resilience on an international scale and its mobilization by a wide range of stakeholders amount to overinvestment: resilience is required to be simultaneously an ideal (perfect) and a universal response (the same for everyone) in ever more complex, open, heterogeneous and uncertain contexts. The delicate transition from theory into practice causes resilience to spill over into politics, which makes it necessary to question the limits and contradictions of some usages and to move the debate toward analyzing discourses on resilience, as well as the challenges of governance. For resilience, the political test will be the moral and ideological assumptions, the contrasts between the discourses held and effective practices, and the debate on the possibility of transforming a retrospective sentence into a prospective tool, of making resilience a guide for future action.

Béné *et al.* put the terms of the debate laconically when they ask "*Resilience: New Utopia or New Tyranny?*" [BEN 12], emphasizing adaptation's "winners and losers".

More broadly, the concept's apparent neutrality draws attention: if elasticity leads to consensus, it tends to erase resilience's inevitably political dimension [REG 12].

In practice, resilience has a dark side because political discourse imposes it as an injunction. It is a divergent intuition, considered alternatively or simultaneously as a process, a condition or a property. To give this intuition a footing in the real world, whatever definition is kept, it is necessary to establish some thresholds and criteria. Hence, the recourse to a norm, discourse, the subjective, etc., because it is necessary to address change and uncertainty: it is necessary to declare resilience. This statement enables a situation, a property, continuities or transformations to be defined *a posteriori* by retrospectively conferring on them a positive, authentic and desirable dimension. Resilience is therefore first a narrative construct, and it is primarily a political discourse.

One of the key issues of resilience is therefore to know who states / declares resilience, what kind of resilience and why, who will win or lose by it, its ideological roots, and see where it all fits.

A critical examination reveals that discourses of resilience are easily instrumentalized to exonerate, justify or legitimize what our societies societies can do worse. Resilience is also an instrument of power that threatens to empty regions and societies of their political issues in the name of consensus, always pushing the people's choice to the background.

Like Janus, resilience is both the beginning and the end, the promise and the threat, the door and the keys. So it is also necessary to accept its dark side: resilience proves to be simultaneously a promising horizon and a major risk.

C.1. Bibliography

[AUG 11] AUGENDRE M., "Risques et catastrophes volcaniques au Japon", in NOVEMBER V., PENELAS M., VIOT P. (eds), *Habiter les territoires à risques*, Presses et Universitaires Romandes, coll. "Espaces et sociétés", Lausanne, pp. 185–206, 2011.

[BEN 12] BÉNÉ C., GODFREY WOOD R., NEWSHAM A. *et al.*, Resilience: new utopia or new tyranny? Reflection about the potentials and limits of the concept of resilience in relation to vulnerability reduction programmes, IDS Working Paper, vol. 12, no. 405, available at http://www.ids.ac.uk/publication/resilience-new-utopia-or-new-tyranny, 2012.

[MES 10] MESCHINET DE RICHEMOND N., REGHEZZA M., "La gestion du risque en France: contre ou avec le territoire?", *Annales de géographie*, vol. 673, pp. 248–267, 2010.

[REG 12] REGHEZZA M., RUFAT S., DJAMENT-TRAN G. *et al.*, "What resilience is not: uses and abuses", *Cybergeo: European Journal of Geography*, 2012, available at http://cybergeo.revues.org/25554.

[URR 13] URRY J., "The complexity turn theory", *Culture & Society*, vol. 22, pp. 1–14, 2005.

[WEL 13] WELSH M., "Resilience and responsibility: governing uncertainty in complex world", *The Geographical Journal*, vol. 180, no. 1, pp. 15–26, 2013.

List of Authors

Bruno BARROCA
Lab'URBA
University Paris-Est Marne-la-Vallée
France

Stéphanie BEUCHER
Lycée Montaigne
Bordeaux
France

Géraldine DJAMENT-TRAN
SAGE
University of Strasbourg
France

Agathe EUZEN
LATTS-CNRS
Paris
France

Julie HERNANDEZ
Tulane University
New Orleans
USA

Claude KERGOMARD
École normale supérieure
Paris
France

Richard LAGANIER
University Paris Diderot Paris 7
France

Antoine LE BLANC
TVES
University Littoral – Côte d'Opale Dunkerque
France

Serge LHOMME
Lab'URBA
University Paris-Est Créteil Val de Marne
France

Céline PIERDET
Université technologique de Compiègne
France

Damienne PROVITOLO
Géoazur
CNRS
Sophia Antipolis
France

Magali REGHEZZA-ZITT
École normale supérieure
Paris
France

Samuel RUFAT
University of Cergy-Pontoise
France

Damien SERRE
RESCUE Solutions
SAS
LIED
University Paris-Diderot Paris 7
France

Index

F, G, H

I, K , L, M

N, O, P

R

S, T

U, V, W

Printed in the United States
By Bookmasters